DATE DUE

Energy Performance of
RESIDENTIAL BUILDINGS

EDITOR: M. Santamouris

Energy Performance of
RESIDENTIAL
BUILDINGS

A practical guide for energy rating and efficiency

First published by James & James/Earthscan in the UK and USA in 2005

ISBN: 1-902916-49-2

Typesetting by S.R. Nova Pvt Ltd., Bangalore
Printed and bound in the UK by The Cromwell Press
Cover design by Paul Cooper Design

For a full list of publications please contact:

James & James/Earthscan
8–12 Camden High Street
London, NW1 0JH, UK
Tel: +44 (0)20 7387 8558
Fax: +44 (0)20 7387 8998
Email: earthinfo@earthscan.co.uk
Web: www.earthscan.co.uk

22883 Quicksilver Drive, Sterling, VA 20166-2012, USA

A catalogue record for this book is available from the British Library

Library of Congress Cataloging-in-Publication Data

Energy performance of residential buildings : a practical guide for energy rating and
efficiency / edited by M. Santamouris.
     p. cm.
  Includes bibliographical references and index.
  ISBN 1-902916-49-2 (hardback)
  1. Dwellings-Energy conservation.   I. Santamouris, M. (Matheos), 1956-
  TJ163.5.D86E528 2005
  697–dc22

                                        2004013404

Printed on elemental chlorine-free paper

Contents

PREFACE

Introduction on the Energy Rating of Buildings

M. SANTAMOURIS

Group Building Environmental Studies, Physics Department, University of Athens, Greece

WHY CARRY OUT ENERGY RATING OF BUILDINGS?

Energy efficiency is a critical issue for high-quality housing. Energy not only represents a high percentage of the running cost of a building but it also has a major effect on the thermal and optical comfort of the occupants.

Recent developments in energy technology make it possible to decrease significantly the energy consumption of buildings, to create housing that is more comfortable and to implement a major decrease in emissions to the environment.

Although the energy efficiency of many home products, (dishwaters, refrigerators, etc.), is available to consumers, the relative energy consumption and efficiency of one of the most expensive investments, their dwelling, which generates the highest energy bills, is not available to the consumer.

Energy rating of a dwelling can provide specific information on the energy consumption and the relative energy efficiency of the building. Energy rating is performed through standard measurements carried out under a specific experimental protocol by specialized and accredited professionals. It is then possible for a potential buyer to have exact information on the energy bills he has to pay, while the owner of a house may be able to identify and pinpoint specific cost-effective improvements.

Energy audits involve specific measurements of the building shell, such as insulation levels and window efficiency, of the lighting and the ventilation, as well as of the heating and cooling systems of the building. The behaviour of the occupants, who explicitly control and affect the internal environment, is also considered. The results are normalized and the building is given a score, for example between 1 and 100, that makes it possible to classify the building against an absolute performance scale.

HISTORICAL DEVELOPMENTS

Energy rating started just after the energy crisis. The concern of the industrialized countries about the high energy consumption of the building sector and, in particular, of residential buildings initiated actions and programmes aiming to rationalize the energy consumption of dwellings.

Following the energy crisis, in 1974 the Swedish government introduced financial support in the form of loans and subsidies for energy-saving measures within the building stock. The aim was to stimulate efficient energy use and to reduce energy for heating. The specific goal was to decrease the gross heating consumption for residential areas by 39–48 TWh over a 10-year period (1978–1988) with a total investment of 31–48 billion SEK (1977 value). The retrofitting measures were voluntary.

In order to evaluate the plan during the initial stages, a programme was set up during the first three years (1977–1980) to assess the characteristics of the building stock and to estimate energy savings. The aim was, on the basis of collected data on energy bills and building technical descriptions etc., to estimate the actual mean energy savings due to several retrofitting methods. In total, 1,144 buildings and apartments were audited, of which 944 were single-family houses and 200 were multi-family buildings.[1]

The continuation of the project[2] focused primarily on experimental and theoretical evaluation of energy conservation measures on statistically selected objects within the Swedish building stock. The results were based on measurements before and after retrofitting during the period 1982–1986 in approximately 300 single- and multi-family houses in seven municipalities.

The Energy Barometer (EB) idea[3] was then developed in Sweden and concerns monitoring the development of building energy use via continuous energy and climatic measurements in houses and reporting changes of building energy over a short time span. The information system was made up of two parts:

- One was aimed at providing for the wider public information on the actual and the predicted energy use of the dwellings, together with analyses of trends and effects of energy-related measures on a large scale. The estimates are based on a representative statistical sample from the population of interest. A sample size of around 1,000 buildings enabled statistically reliable monitoring of how the Swedish national use of energy has varied over time.

- The second part aimed mainly at providing individual house owners with a means of monitoring their energy bills. Occupants of a building connected to the Energy

Barometer could analyse their own energy situation and also view it in relation to that of the population as a whole.

In the USA, energy rating systems have been used since the 1980s. However, the idea of using Energy Efficiency Mortgages (EEM) associated with home energy rating systems, which has been applied since 1990, has considerably increased the penetration of rating systems in the residential market. The existence of an energy audit has substantially helped the mortgage industry to make loans for energy improvements. In 1992, the US Congress voted through the Energy Policy Act, which provided for the establishment of credible voluntary national guidelines for home energy rating schemes.[4] Then a Home Energy Rating System (HERS) program was started in five pilot states, Alaska, Arkansas, California, Vermont and Virginia. Today 31 states have adopted the MEC 1993 version or its equivalent. However, only 2% of new homes actually receive an energy rating in the USA, and most are utility programmes with tax-payer subsidies.

In Denmark, energy rating schemes have been prepared for large commercial buildings since 1992. The schemes were extended to residential buildings, a year later, 1993. The scheme is mandatory, relatively costly and quite comprehensive.

In the UK, in the 1980s and 1990s Building Research Establishment (BRE) performed hundreds of multi-year energy audits in residential buildings. From these, BRE was able to develop the Domestic Energy Model (BREDEM). Today three labelling schemes are in use:

- the Standard Assessment Procedure (SAP) on a scale from 1 to 120, which has been required by building regulations for new housing since 1995
- the National Home Energy Rating, (NHER), on a scale of 0 to 10
- an estimate of the carbon dioxide emitted each year as a result of a home's energy use (the carbon index).

The principal target for labelling in the UK is social housing and today almost 75% of the social rental housing has been labelled, while about 20% of owner-occupied housing has been labelled at the time of sale.

In Ireland, the National Irish Centre for Energy Rating (NICER) created the Energy Rating Bench Mark (ERBM) in 1992 to deal with existing buildings. Almost 8,000 new houses per year are labelled with ERBM.

In Spain, and in particular in the Basque Country, an energy rating system has been developed to classify new residential buildings. Energy-efficiency certificates are awarded to buildings in two stages. A first certificate is awarded in the design phase and a final one is provided to the finished building.

Finally, in the Netherlands, a rating scheme, EPB, was developed in the mid 1990s. The method mainly targets social housing. A new method, EPA, was put into operation during the year 2000.

PROBLEMS AND FUTURE PROSPECTS

Energy rating of dwellings has reached a high level of scientific maturity, in particular in countries where energy rating has been used for many years. It is expected that the new European Directive for Energy Efficient Buildings will enforce the use of such methodologies and will expand the application of home rating systems in all European countries and, in particular, in Southern Europe. The new Directive is asking for an energy rating of many types of buildings and brings energy certification into everyday life. Member States have to prepare their national methodologies, but an effort is made to homogenize the rating techniques to be used as much as possible.

However, a number of barriers to the widespread use of home rating systems have been identified in many countries. The main barriers appear to include:

- lack of owners' awareness of energy efficiency benefits
- insufficient awareness and training of property managers, builders and engineers
- low energy costs for both electricity and thermal energy
- lack of sufficient funding to assist the penetration of home rating systems in the real market
- the relatively high cost of home energy systems
- lack of data on the energy consumption of dwellings, at least in many European countries, which do not permit comparisons to be made and defaults defined
- lack of specialized professionals to perform energy audits and ratings in residential buildings
- lack of builder incentives
- lack of financial interest and lack of financial gains for the owners, the builders and the real-estate managers.

During the last few years, environmental issues other than energy have been considered in the assessment of buildings.[5] Ecological parameters such as shortage of raw materials, water consumption, indoor air quality, noise and pollution, health aspects and waste treatment are the main considerations. It is believed that environmental rating systems, involving energy issues and life-cycle analysis, will have a very fast development in the near future.

THE EUROCLASS METHOD

As described above, several energy rating techniques have been proposed and are in use. Each methodology has to be based on an experimental protocol for collecting energy data, a theoretical algorithm to normalize the energy consumption and an algorithm to classify buildings. It is thus very reasonable for each national methodology to be adapted to the characteristics of the national building stock, the national methodology for measuring energy and the specific climatic characteristics of the country. For example:

- Cooling has only very recently become important for the residential sector. Thus, most of the existing rating methodologies do not consider at all the specific energy consumption for air conditioning. Given the general energy tendency in Europe and the requests of the new European Directive, a new European classification methodology must involve a cooling methodology.

- In Nordic countries solar gains do not significantly influence the heat balance of the building and thus a normalization approach based on degree days seems very reasonable. However, such an approach can be used in Southern countries, where solar and internal gains may play a very important role.
- The characteristics of residential buildings differ considerably from state to state. Different types of materials are used and different construction techniques are employed, while the mode of operation is sometimes completely different. Thus, the benefit of a utilization factor is doubtful as long as generalization of data from numerous buildings in a state makes the results for one specific building uncertain. At the same time it does not allow the use of the method in another state.
- Billing of the energy consumption differs between the various countries. A methodology to calculate the specific energy consumption of a building based on energy bills is highly influenced by the billing methodology and thus national methodologies need to be fully adapted to the specific conditions.
- The energy consumption of the residential sector varies considerably between states. Classification of buildings based on their energy consumption is always based on a large database of national energy data and thus the divisions between the classes have a very strict national character.

EUROCLASS is a new methodology recently developed through the European SAVE program and aims to overcome most of the above problems. EUROCLASS has run between 1998–2000. The method proposes experimental protocols that may be used in and adapted to each state, while it also proposes a theoretical method that includes all specific energy uses and treats energy normalization in a very innovative way. The proposed methodology seems to provide answers to most of the problems of the existing methodologies and thus may be the platform for future developments in the field.

DESCRIPTION OF THIS BOOK – HOW TO READ THE BOOK

The aim of this book is to inform readers about the latest developments in the field of energy rating of dwellings. In particular, the book presents the new European experimental and theoretical methodology for assessing and rating the energy efficiency and performance of dwellings.

The book is a collaborative work undertaken by the University of Athens, Greece, the Belgium Building Research Institute (BBRI), the Royal Technical University of Stockholm and the University of Seville in Spain, in the frame of the European SAVE program, EUROCLASS.[6] The project was coordinated by the Group of Building Environmental Studies of the University of Athens, Greece.

The overall book is made up of five chapters plus three Appendices. The methodology has been prepared with the active participation of all partners in all parts of the work. The authors mentioned in each chapter have undertaken to report the specific work carried out by the whole consortium.

Chapter 1 deals with the presentation of some selected theoretical and experimental methodologies proposed and applied for the energy characterization and rating of dwellings.

Chapter 2 deals with the presentation of the proposed experimental protocols – methods for the energy characterization of dwellings. Two separate protocols have been designed to fulfil the same requirements, but with different degrees of detail. The costs, the resources, the delivered information (service) and the level of accuracy of the results will allow the customer (occupants) to choose which protocol to use. The results generated by each protocol are compatible in rating since both have a similar experimental platform.

The two protocols are called:

- Billed Energy Protocol (BEP)
- Monitored Energy Protocol (MEP).

Chapter 3 deals with normalization techniques. In a general way, normalization takes into account:

- the size of the building
- the climate of the external environment
- the climate of the internal environment.

The main existing normalization techniques for space heating and space cooling are presented and their limitations and tolerances are discussed in detail. A new proposed normalization method, the Climate Severity Index, is then presented. The method presents many advantages compared with the classical normalization techniques.

Chapter 4 describes the software developed within the frame of the EUROCLASS project, which is designed to apply the proposed rating methodology for dwellings. EUROTARGET, which is the name of the software, is described in detail and examples are given. A full copy of the software is enclosed with the book.

The methodology presented in the above chapters has been applied in four countries (Belgium, Greece, Spain and Sweden). Chapter 5 describes the results of the procedure, the problems encountered and the solutions envisaged to solve these problems.

Chapter 5 is divided into two parts. The first part deals with the results of the BEP procedure, while the second part deals with the results of the MEP.

The two protocols have each been applied in the four considered countries. The application of the methodology, the way to carry out the monitoring, the choice of normalization techniques and the calculations have been implemented at a national level.

Finally, Appendix 1 presents the audit forms prepared and used in the frame of the Euroclass project. The forms should be very useful for anyone interested in applying the proposed rating methodology.

REFERENCES

1. Bostadsdepartementet, 1980, *Energispareffekter i Bostadshus där Åtgärder Genomförts med Statligt Energisparstöd*. Expertbilaga 5 till SOU 1980:43 – Program för energihushållning i befintlig bebyggelse. Ds Bo 1980:8.

2. Elmroth A, Hjalmarsson C, Norlén U, Rolén C, *et al.*, 1989, *Effekter av energisparåtgärder i bostadshus*, Rapport R107:1989, Byggforskningsrådet, Stockholm, Sweden.

3. Norlen U, 1985, 'Monitoring energy consumption in the Swedish building stock', *Proceedings of Conference on Optimisation of Heating Consumption, Prague*, Swedish Institute for Building Research, Gävle.

4. Fahrar B C, 2000, *Pilot Program Report: Home Energy Rating Systems and Energy – Efficient Mortgages*, NREL/TP–550-27722. Available through NREL.

5. *Environmental Assessment of Buildings*, 1995, A Thermie Programme Action, No B 108, European Commission, Directorate General for Energy, Brussels.

6. Santamouris M (editor), 2001, *Final Report of the EUROCLASS Project*, SAVE program, European Commission, Directorate General for Transport and Energy, Brussels.

CHAPTER 1

Review of selected theoretical and experimental techniques for energy characterization of buildings

P. WOUTERS AND X. LONCOUR

Division of Building Physics and Indoor Climate, Belgian Building Research Institute, Brussels

INTRODUCTION

The aim of this chapter is to present selected theoretical and experimental energy rating methodologies developed for dwellings. It is not the intention of the authors to present all existing methodologies, but to inform the readers about the main types of proposed theoretical methodologies and to present representative methods of each type.

This chapter consists of two parts:

- The first part presents the result of the bibliographic study on the state of the art of the theoretical and experimental techniques used to determine the energy consumption of residential buildings.
- The second part deals with the in situ identification of the transmission heat loss coefficient + (UA value) and the gA values (characterizing the solar gains) of buildings.

Part 1. State of the art – measurement techniques

CALCULATION METHODS AND EXPERIMENTAL TECHNIQUES

Several remarks need to be made before going into more detail:

- Much information can be found here about the *calculation methods* used to determine the energy consumption of the buildings.
- Some of these calculation methods are firmly based on experimental data. It should be noted that very often the articles focus more on the description of the calculation method than on the way of realizing the experimental monitoring.
- In general, few papers focus on the experimental techniques used to collect the data on site.
- Some of the experimental techniques described in the literature are quite old (some of them were developed more than 20 years ago). These experimental techniques do not take into account the latest developments in

measuring techniques (for instance, the use of the Internet) and are therefore less interesting within the context of this project.

We mention in this chapter the different calculation methods. We give the references of the papers concerned and, when they are available, we describe the techniques used to collect the relevant information on site.

The different calculation methods that are considered in this document can be regrouped into five categories:

1. the university projects and the Energy Barometer
2. the Save HELP method
3. short-term energy monitoring and primary and secondary term renormalization
4. neural networks
5. other related methods.

The university projects and the Energy Barometer

THE UNIVERSITY PROJECTS

The two 'university projects' have been developed in Sweden. References [1] and [2] describe these two projects. Since these references are in Swedish, we give more detail here about the methodology applied in the scope of these two projects than for other methods described in this document.

In 1974, the Swedish government introduced financial support in the form of loans and subsidies for energy-saving measures within the existing building stock. The aim was to stimulate efficient energy use and to reduce energy for heating. The goal was to decrease the gross heating consumption for residential areas by 39–48 TWh over a 10-year period (1978–1988) with a total investment of SEK 31–48 billion (1977 value). The retrofitting measures were voluntary.

In order to evaluate the plan during its initial stages, a programme was set up during the first three years (1977–1980) to determine whether or not it was worth continuing the plan. The evaluation was also intended to assess the characteristics of the building stock and to estimate energy savings. The aim was to estimate, on the basis of collected data on energy bills and building technical descriptions, the

Table 1.1 The efficiency of heating systems during winter and summer. Values are given for the various types of single- and multi-family dwellings

Efficiency	Electrical heating		Oil/gas		District	
	Single-family	Multi-family	Single-family	Multi-family	Single-family	Multi-family
η_{winter}	1.00	1.00	0.65	0.85	0.95	0.95
η_{summer}	1.00	1.00	0.30	0.65	0.95	0.95

actual mean energy savings resulting from several retrofitting methods. Reference [1] describes this project, which is called the 'university project' (UP1), and the method applied.

The task

The task was to calculate the energy savings resulting from different retrofitting means on the basis of collected energy-bill data and of an inspection of building technology and the systems installed. In total, 1,144 buildings and apartments were audited, of which 944 were single-family houses and 200 were multi-family buildings. Broadly, the following steps were taken:

- Energy bills were collected before and after the retrofitting was carried out.
- Bill values were normalized according to temperature variations over the years.
- The energy saved was measured in corresponding litres of oil, during a reference year.

It was found that 841 buildings had a complete set of bills and 303 had an incomplete set. More details are given below, especially on the underlying assumptions made.

The static energy signature was determined for each building, and from this data could be statistically applied to the Swedish building stock as a whole.

The indoor temperature

The indoor temperature was assumed to be 21°C, with the exception of the period during and shortly after the oil crisis. The motivation for this was that information on indoor temperature that was supplied by the inhabitants was considered to be inaccurate.

The outdoor temperature

The Swedish Meteorological and Hydrological Institute (SMHI) supplied outdoor temperatures on the basis of data from climate stations across the country. Surveyors had to judge which climate station best described the temperature at a particular site. This eliminated local variations of temperatures.

The efficiency of the heating system

The efficiency of heating systems varied from building to building, but within this work was assumed to be the same for various categories of houses. The values decided upon are listed in Table 1.1.

Energy saving calculations were made as if all buildings were heated with oil and a conversion factor was used for each fuel used as a heat source, based on the energy content of the fuel (Table 1.2).

Table 1.2 Fuel conversion factors

Fuel (unit)	Conversion factor
Oil (m³)	1.000
Gas (m³)	0.471
Electricity (kWh)	0.101
District heating (kWh)	0.101

Table 1.3 Values assumed for energy consumption by appliances and hot water

Energy (kWh)	Single-family house	Multi-family house
Electrical appliances	4,600	2,800
Hot water	4,000	3,500

Energy for appliances and hot water

The use of energy for electrical appliances and hot water was assumed to be as shown in Table 1.3 for each household.

Degree-hours

Among the climatic factors, such as temperature, solar irradiation, wind, snow, long-wave irradiation and moisture, that influence the heat balance of a building, the outdoor temperature was considered to be the most important factor. Because of the seasonal variations in outdoor temperature, the periods taken into consideration before and after retrofitting were chosen so as to be complete years, if possible. The number of degree-hours was computed from

$$Q = T(\theta_i - \theta_e) \quad (°C \cdot h)$$

where T is the number of hours during the heating season, θ_i is the mean indoor temperature and θ_e is the mean outdoor temperature based on mean monthly values. Linear interpolation was used to calculate the mean outdoor temperature.

For the calculations, the heating season was limited to months with sufficiently low outdoor temperature by starting in October and ending in April.

In order to compute the number of degree hours for a reference year, the collected outdoor temperatures for the years 1972–1979 were used.

U values

U values were calculated in accordance with norms. For double- and triple-glazed windows, the values were 3.0 and 2.0 W/(m² · °C), respectively. The glazed part was assumed to be 70% of the gross window area.

Calculation model

The calculation model was based on an energy balance for the building, which was split up into a summer season and a

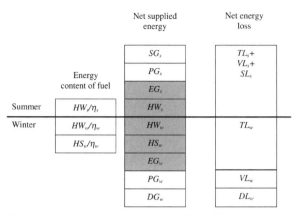

	Energy content of fuel	Net supplied energy	Net energy loss
		SG_s	TL_s+ VL_s+ SL_s
		PG_s	
		EG_s	
Summer	HW_s/η_s	HW_s	
Winter	HW_w/η_w	HW_w	TL_w
	HS_w/η_w	HS_w	
		EG_w	
		PG_w	VL_w
		DG_w	DL_w

Figure 1.1 Net energy balance for a building. The shaded parts of the net supplied energy depict the energy supplied for an electrically heated building

winter season. Figure 1.1 illustrates the various energy parts that make up the whole.

In Figure 1.1 and the equations below, the following abbreviations are used:

SG = solar gains
PG = heat from people
EG = heat from household appliances (electrical gains)
HW = net heat supplied for hot water
HS = net heat supplied by the heating system
TL = heat loss due to transmission
VL = heat loss due to ventilation
DL = heat loss via drainage

For electrically heated buildings, the energy consumption is

$$W = TL_w + VL_w + DL_w - PG_w - SG_w + HW_s + EG_s$$

For oil-, gas- or district-heated buildings, the energy consumption is

$$W = (TL_w + VL_w + DL_w - PG_w - SG_w - EG_w)/\eta_w$$
$$+ HW_s/\eta_s$$

If it is assumed that the losses due to transmission and ventilation are proportional to the amount of degree-days, that the drainage loss, the heat gains from people and solar irradiation and household appliances are proportional to the number of hours in the heating season (winter) and that the household appliance gains during the non-heating season are proportional to the number of hours during the non-heating season, the two equations can be rewritten in the form

$$W \cdot H = b \cdot Q + c \cdot T + d \cdot P$$

where
W = the collected energy consumption
H = conversion factor
Q = the number of degree-hours during the period considered
T = the number of hours during the heating season
P = the number of hours during the non-heating season
b = the building heat-loss factor, litres of oil per hour per degree

c = the winter factor, litres of oil per hour
d = the summer factor, litres of oil per hour

The two first terms on the right-hand side of this equation describe the energy consumption during winter, while the last is for the summer period. The heat-loss factor b is a measure of the insulation and of the amount of ventilation of the building. The summer factor d describes the energy consumption during the non-heating season (summer), whereas the winter factor c is more difficult to interpret. All three factors depend on the efficiency of the heating system. The factors are determined by means of linear regression on data from buildings with extensive information on energy consumption before and after retrofitting.

With the assumption that the winter factor c and the summer factor d are known and remain unchanged during the period before and after retrofitting, the heat-loss factors before retrofitting, b_b, and after retrofitting, b_a, are

$$b_b = \frac{1}{Q_b}(W_b \cdot H - c \cdot T_b - d \cdot T_b)$$

and

$$b_a = \frac{1}{Q_a}(W_a \cdot H - c \cdot T_a - d \cdot T_a)$$

Now all these values can be determined and can be used to calculate the energy consumption that would have been measured during a year with 'normal' climate:

$$(W \cdot H)_{b,r} = b_b \cdot Q_r + c \cdot T_r + d \cdot P_r$$

and

$$(W \cdot H)_{a,r} = b_a \cdot Q_r + c \cdot T_r + d \cdot P_r$$

where index r denotes the reference year.

The saving of energy due to retrofitting is the difference between $(W \cdot H)_{b,r}$ and $(W \cdot H)_{a,r}$. First, the net energy has to be calculated with the efficiency of the actual heating system. Then, divided by the efficiency of an oil-based boiler, the saving expressed in litres of oil per year is found to be

$$S = (b_b - b_a)Q_r \frac{\eta_w}{\eta_{w\,oil}}$$

where
S = the energy saving if the building were heated using oil (litres/year)
b_b = the heat-loss factor before retrofitting (litres/degree per hour)
b_a = the heat-loss factor after retrofitting (litres/degree per hour)
Q_r = the number of degree-hours during a reference year
η_w = the efficiency of the actual heating system during winter
$\eta_{w\,oil}$ = the efficiency of the oil-based heating system

The second 'university project' (UP2) described in reference [2] focuses primarily on the experimental and theoretical evaluation of energy conservation measures on statistically selected objects within the Swedish building

Table 1.4 Correspondence between measured and theoretically predicted savings

Retrofit	Building type	Saving in relation to energy consumption prior to retrofit	
		Measured (%)	Predicted (%)
Triple glazing	Single-family	6	10
Retrofit package	Single-family	19	25
Conversion to electric heating	Single-family	22	20
Triple glazing	Multi-family	9	10
Additional insulation of attic	Multi-family	5	6
Regulating package	Multi-family	4	3
Retrofit package	Multi-family	14	16
District heating	Multi-family	24	15

stock. The results were based on measurements before and after retrofitting during the period 1982–1986 in approximately 300 single- and multi-family houses in seven municipalities. The most important results were:

- In most cases, energy conservation measures gave statistically established savings. In many cases, however, savings were small.
- For most cases, the measured savings agreed with the theoretically predicted savings (Table 1.4).
- Energy savings of heat pump installations in single-family houses were 40–60% of pre-retrofit energy consumption.
- The temperature decreases expected as a result of some retrofits were not achieved.

Background

UP2 was a continuation of UP1, but one of the aims was to improve the accuracy of the results where the procedures of UP1 were considered to be weak. Monitoring indoor and outdoor temperatures led to improvements, as did the use of energy meters, which were read on a weekly basis. In addition, oil burners were supplied with run-time sensors. Two models were used in the analysis: the measurement model and the theoretical model. Common to both is that the most important parameters of the energy performance of the building are considered.

Energy consumption is presented in kWh per apartment per normal year (here, energy consumption is that required for space and domestic water heating during one year). The climate is normalized to a year where:

- the outdoor temperature is equal to the mean temperature during the years 1951–1979
- the solar irradiation is equal to the mean solar irradiation during the years 1955–1979

In the measurement model, series of weekly collected energy consumption data were used. In the theoretical model, the descriptive technical information was collected and used, together with default values, to determine the energy

performance parameters. For both models, the following data are required:

- parameters that describe the energy performance of the building
- the indoor temperature
- climatic parameters that describe the climate of the normalized year.

The energy saving is the difference between the energy consumption before and after retrofit. The energy saving from the measurement model is called the measured energy saving, whereas the value calculated from the theoretical model is called the expected energy saving. The energy saving is dependent upon which indoor temperature is used. A way of defining the energy saving is as follows:

$$[\text{energy saving}] = [\text{effect of retrofit}]$$
$$+ \left[\begin{array}{l} \text{effect of changed} \\ \text{indoor temperature} \end{array} \right]$$
$$+ \left[\begin{array}{l} \text{effect of other} \\ \text{influencing factors} \end{array} \right]$$

The analysis was set up so as to minimize the last term (i.e. to make it negligible). If this is done, the following applies:

- The measured energy saving: any change in indoor temperature is represented in this change such that

$$\left[\begin{array}{l} \text{energy saving in} \\ \text{measured temperature} \end{array} \right] = [\text{effect of retrofit}]$$
$$+ \left[\begin{array}{l} \text{effect of changed} \\ \text{indoor temperature} \end{array} \right]$$

- The expected energy saving: the indoor temperature before and after retrofitting has standardized values that are assumed to give the same thermal comfort.
- Other types of energy savings: with the two models (measurement and theoretical), and two temperatures (measured and standardized), four different energy saving variants can be calculated for every building. The measured energy savings (with measured indoor temperature) and the expected savings (with standardized indoor temperature) are those that are mostly used in this report.

Measurement model for energy consumption

Calculations are made according to

$$W = b \cdot Q + c \cdot T + d \cdot P \text{ (kWh/apartment/year)}$$

where

W = the annual energy consumption per apartment per year; in certain cases this is converted to kWh by means of conversion factors that correspond to the energy content of the fuel (kWh/apartment/year)

Q = the number of degree hours during the heating season (°C · h/year)

T = the number of hours during the heating season (h/year)

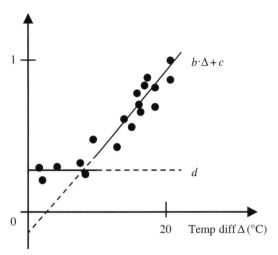

Figure 1.2 The relationship between the energy consumption per apartment per unit of time and the difference between indoor and outdoor temperatures, Δ

P = the number of hours during the summer season (h/year)

b = the building heat-loss factor (kWh/apartment/°C · h)

c = the winter factor, which quantifies temperature-independent energy losses (ground, drainage) and heat gains (household appliances, persons, insolation) (kWh/apartment · h)

d = the summer factor, which quantifies the mean energy for heating of water during summer (kWh/apartment/h)

Figure 1.2 shows the relationship between the energy consumption per apartment per unit of time and the difference between indoor and outdoor temperatures, Δ.

In the model, values for b and c are estimated from measured values by the least-squares method, such that

$$b = \frac{\sum\limits_{i=1}^{n} W_i(\Delta_i - \overline{\Delta})}{\sum\limits_{i=1}^{n} (\Delta_i - \overline{\Delta})^2} \quad \text{and} \quad c = \overline{W} - b \cdot \overline{\Delta}$$

where

W_i = the weekly mean energy consumption during period number i divided by the length of the period in hours

Δ_i = the mean temperature difference (indoor − outdoor) during period i

$\overline{W}$ and $\overline{\Delta}$ are the mean values for the whole measurement period in question. The summer factor d has the same standardized value for houses of the same type.

Alternative measurement models have been analysed, for example one with a third energy parameter that represents insolation. No major differences in the results for the heating season were noted between the three-parameter model and the two-parameter model. Some small differences were explained by correlation between the week-wise solar irradiation and the outdoor temperature and by the slight influence of the solar irradiation on energy consumption during the periods when measurements were made.

From 50 climate stations across the country, daily mean outdoor temperatures were obtained. For a particular building, there is often a systematic difference between the outdoor temperature at the building and the chosen climate station, usually the nearest to the building site. A correction was made for this difference between the measured mean temperature at the building and that at the climate station during the measurement period.

The theoretical model for energy consumption

If an energy balance model with access to outdoor climate data is set up, the annual energy consumption can be calculated before and after a retrofit. The calculated energy consumption for space and water heating is found by setting the supplied energy equal to the energy loss. The time period of the heating season is set to be the same as in the measured case.

Certain assumptions were made concerning the energy for household appliances, the energy for heating water, the energy from persons and the gains from insolation (see UP1 and below).

The supplied net energy (i.e. the paid-for energy) for heating space and water can be formulated as (see above for meaning of the notation)

$$W_{net} = HS_w + HW_w + HW_s = (TL_w + VL_w)$$
$$+ (DL_w - EG_w - PG_w) - SG + HW_s$$

which is rewritten as

$$W_{net} = b(\theta_i - \theta_e) \cdot T \cdot \eta_w + c \cdot T \cdot \eta_w - f \cdot I \cdot T \cdot \eta_w$$
$$+ d \cdot P \cdot \eta_s$$

with the introduction of two new parameters:

f = the solar area, which is a fictitious window that transmits all insolation (m²/apartment)

I = the mean solar irradiation during the heating season (kWh/m²/h/year)

The following assumptions have been made:

- The sum of the transmission and ventilation losses is proportional to the number of degree hours.
- Drainage losses, the heat from household appliances and the heat from people are proportional to the length of the heating season.
- The energy for water heating during the warm season is proportional to the length of the season.

The gross energy consumption is therefore

$$W_{gross} = [b(\theta_i - \theta_e) + c - f \cdot I] \cdot T + d \cdot P$$

The heat-loss factor b is calculated as the sum of transmission loss factor for the envelope components (the $U \cdot A$ value), and the ventilation loss factor ($0.33 \cdot n \cdot V$ with n representing the rate of air change in volume V).

The loss factor can therefore be expressed as

$$b = 0.001 \frac{U \cdot A + 0.33 \cdot n \cdot V}{\eta_w} \text{ (kW/apartment/K)}$$

The winter factor c and the summer factor d are given default values.

The solar w aperture is calculated with four reduction factors and the gross window area A_w, such that

$$f = A_w \frac{f_s \cdot f_f \cdot f_c \cdot f_t}{\eta_w} \text{ (m}^2)$$

with

f_s = the shading factor
f_f = the frame factor
f_c = the curtain factor
f_t = the transmittance factor

The Energy Barometer

The idea of the Energy Barometer (EB) is to monitor the development of building energy use via continuous energy and climatic measurements in houses and to report changes of building energy over a short time span. Analyses will be presented on line. The word 'barometer' is intended to refer to measuring the pressure on the energy market. The information system is made up of two parts. The first part is aimed at the population level, providing estimates of actual and predicted energy use, together with analyses of trends and the effects of energy-related measures on a large scale. The estimates are based on a representative statistical sample from the population of interest. A sample size of around 1,000 buildings makes possible a statistically reliable monitoring of how the Swedish national use of energy varies across time.

The second part is aimed at providing individual house owners with a means of monitoring their energy cost budgets. Buildings connected to the EB will not only be able to analyse their own energy situation but also to see it in relation to the population as a whole.

The Energy Barometer has been developed to measure energy use in relation to internal and external climates in different types of single-family house during periods with and without heating. It is also expected to provide a basis for analysis and evaluation of energy efficiency measures. The electrical energy use by household appliances and the energy use for heating, including heating of tap water and indoor temperature, are measured in each selected building. The whole theoretical background is described in references [3–5].

The calculation principles

Two kinds of model can be used, based on either a static or a dynamic energy balance. The background relating to the static heating balance continues here.

Consider a whole year consisting of T hours when the house is heated ('winter hours') and P hours when the house is not heated ('summer hours'). Let $\Delta\theta$ be the average indoor–outdoor temperature difference (°C) during this period and let I be the global solar radiation on a horizontal surface during the heating period (kWh/m^2). Then the following expression can be used to calculate the annual energy use, E_i(kWh), for the heating in house i:

$$E_i = c_i T_i + b_i T_i \Delta\theta + f_i I_i + d_i P_i$$

The energy parameters c, b, f and d have specific values for each house. The climatic variable variables T, P and I depend on the climatic conditions for the region where the house is situated. The number of degree-hours, $Q = T\Delta\theta$, differs between houses depending on the temperature conditions and the length of the heating season. The variables T, P and I may to a first approximation be given the same values for all houses in a particular region.

For the purpose of estimating total energy for heating, it is sufficient to include in the model solar irradiation on a horizontal surface.[6] When the end points of the heating season are known, the contribution from the solar irradiation during the heating season of a normal year can be estimated using a sinusoidal approximation.

The energy consumption is determined on the basis of a static energy balance. The purpose is to estimate the 'normal' annual energy use for heating (space and tap water). Under stationary conditions, heat will be stored in or released from the house over a specific period of time. This assumption enables the formulation of a linear static heat balance equation for the house – these are called energy signature models.[7]

By dividing the previous equation by T, we obtain the average energy use per unit time w (kWh/h) during the heating season. The constant c is the average energy use per unit time when there is no temperature difference and no solar irradiation. ε is the error term:

$$w = c + d\Delta\theta + fs + \varepsilon$$

The model assumptions are valid provided that the time period chosen is long enough for the heat stored in or released from the building to be very small compared to the total energy supplied during the period.

Two reasons for using a dynamic energy balance model instead of a static model of the type discussed above are:

- Energy use in buildings is a dynamic process, e.g. the thermal inertia delays the 'response' of the building to changes in the outdoor climate.
- Information in the data is lost in the aggregation from the hourly data to the daily or weekly data needed for the static model.

A general dynamic model can be written as an equation that expresses the energy use $w(t)$ during hour number t as a linear function of the following variables at hour t and the preceding hours $(t-1)$, $(t-2)$, . . .

- the energy use $w(t-1)$, $w(t-2)$, . . .
- the indoor temperature θ^{in} and the indoor temperature $\theta^{in}(t-1)$, $\theta^{in}(t-2)$, . . .

- the outdoor temperature θ^{out} and the outdoor temperature $\theta^{out}(t-1), \theta^{out}(t-2), \ldots$
- the solar irradiation $I(t)$ at hour t and the solar irradiation $I(t-1), I(t-1), \ldots$

The following ARX model (AutoRegressive model with eXogeneous inputs) with 11 free parameters has been used:

$$w(t) + a_1 w(t-1) + a_2 w(t-2)$$
$$= c_1 + b_1 \theta^{in}(t) + b_2 \theta^{in}(t-1) + b_3 \theta^{in}(t-2)$$
$$+ b_4 \theta^{out}(t) + b_5 \theta^{out}(t-1) + b_6 \theta^{out}(t-2)$$
$$+ f_1 I(t) + f_2 I(t-1) + f_3 I(t-2) + \varepsilon(t)$$

Reference [7] also describes the way to generalize the results obtained for the houses investigated, the selection of the houses, the estimation of the total energy use for heating and of the annual change, the calculation of the error in the estimates and the sample size.

The Energy Barometer system gives a solution to the combined problem of obtaining (i) timely and (ii) reliable estimates of end-use consumption.

- **Timely estimates.** Statistics about building energy end-use have hitherto often been based on postal questionnaires with obvious problems of measurement quality and have been reported more than one year after the energy use occurred. With the method presented, using Internet and modem communication systems, data can be collected continuously and the quality of the measurements can be specified and controlled. In addition, climatic corrections can be made to the estimated annual energy use.
- **Reliable estimates.** The reliability of the estimates depends on statistical procedures and the assumption that the selected houses represent the population of single-family houses. The results obtained indicate that a time step of 24 hours is too short for the dynamic properties of a house to be averaged out and/or that the houses do not behave according to a model with a time-constant parameter system. This means either that static models should be based on weekly data or that models with time-varying parameter systems should be tested. The recommended length for the observation period is about 7 to 10 weeks.

Reference [8] describes the way in which houses equipped with heat pumps can be taken into account in the Energy Barometer system.

A first application of the method has been realized, although not yet totally at the full scale; the subproject is called the 'Virtual Housing Laboratory' (VHL).[9] This is a system for simulating total energy use in the single-family housing stock. The system is based on a sample with very detailed data for 737 single-family houses and actual climate data from nearby weather stations. The VHL is part of the Energy Barometer project, which in its full version implies collecting energy and climate data on an hourly basis for about 1,000 sampled houses.

THE TECHNICAL SOLUTION USED TO COLLECT THE DATA

The specification of the type of sensors used and the accuracy of measurement are presented in Table 1.5. Reference [10] describes the specifications of the sensors used for a similar type of investigation.

The description of the measurement protocol and of the material used to collect the data is summarized in reference [11]. This article presents the concept of the Energy Barometer used in Sweden with an emphasis on the data handling system and the associated communication infrastructure. We summarize here the main characteristics of the system.

The data handling system is based on a communication device, a residential gateway ('E-box') constructed by Ericsson Radio Systems. The practical reasons for using this system were:

- Intrusion into the home environment should cause minimal inconvenience for the residents in terms of, for example, wiring and extensive installation work.
- The system should be versatile and open enough to allow other uses of the installed communication facilities and hence decrease the part of the cost related to the measurements necessary for this project.
- The system allows remote operation and maintenance to avoid travel costs etc.

The E-box is basically a computer, the operating system of which is specifically designed to allow a number of different applications to run in parallel. Practical information and the specifications of the E-box system can be found on the Ericsson's website.[12]

In each house the sensors described above are installed. They are connected via a network to the E-box. LonWorks has been used as local network since it enables physical-layer communication on the 220 volt cables already present in the houses. The sensors are then simply plugged into the

Table 1.5 Details of the sensors used for measuring different variables and their accuracy

Variable	Unit	Accuracy	Sensors and additional units
Energy			
Electricity (electric heating, water-borne heating, heat pump)	kWh	±2%	kWh meter with output of pulses
District heating	kWh	±2%	Integration meter with output of pulses
Oil burner	Litre	±5%	Flow meter with output of pulses
Run time: oil	h	±1%/h	Sensor or electromagnetic sound
Temperature: indoor, outdoor	°C	±0.3°C	Temperature sensor

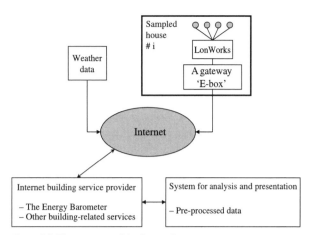

Figure 1.3 The structure of the Energy Barometer system

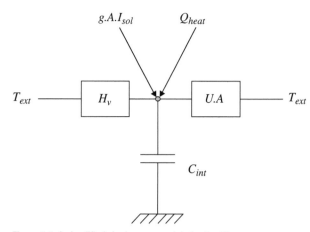

Figure 1.4 A simplified single-zone model of a dwelling

wall and no extra wiring is necessary. Figure 1.3 shows the structure of the EB system. Raw energy and temperature data are collected continuously and stored in each local E-box and transferred to a central server on a diurnal basis via the Internet. Weather data are sent from the nearest official climate station.

From each house the following data have to be collected:

- the hourly energy use for heating (space and tap water)
- the hourly energy use of energy for household appliances
- the hourly average indoor temperature.

In addition to these data, technical data on, for example, occupancy, heated area, year of construction, are also needed. This type of data is collected when the house is inspected in connection with installation of sensors and the E-box. Reference [13] gives a sample inspection protocol.

THE SAVE HOUSE ENERGY LABELLING PROCEDURE (HELP)

This procedure has been developed in the framework of the EU-financed Save HELP project.

Theoretical background

Reference [14] describes the theoretical method used.

The first objective was to characterize the energy performance of *non-occupied buildings* and the following procedure was developed:

1. Light monitoring of the studied dwelling according to a standardized test procedure.
2. The collected data set is then used to identify a simplified single-zone thermal model (as presented in Figure 1.4, a global *UA* value, *gA* value and possibly C_{int}, the internal capacitance).
3. Once the model is obtained, a simulation can be carried out with defined climatic and internal-gains data sets in order to determine a standardized energy consumption level, which can be used as an energy performance index for the considered building.

Since the use of light measurement techniques and methods is required for large-scale building certification, it has been chosen not to apply co-heating, nor to control the surrounding ambiences such as the climate in the basement or the attic.

The heated space is treated as a single zone (internal doors are assumed to be open).

DETERMINATION OF A TEST PROCEDURE

A constant indoor temperature strategy is applied. The advantages are:

- It is easier to obtain the homogeneity of indoor conditions.
- It is applicable to all types of dwellings where energy consumption is easily measurable (gas- and electricity-heated houses).

The disadvantages are:

- Determination of the internal capacity is almost impossible.
- The accuracy of the heating power measurement is reduced; there is no continuous measurement of the consumption.

THE CHOICE OF MEASURED SIGNALS

The sensors and measurements were as follows:

- a few temperature sensors in the heated zone
- one measurement of the outdoor temperature
- one extra temperature sensor in each unheated zone (basement, attic...); not used for analysis but for checking the test conditions
- measurement of the air change rate with PerFluorTracergas (PFT)
- measurement of solar radiation with photovoltaic (PV) elements
- periodic measurement of the energy consumption (if possible, on a daily basis).

THE METHOD OF ANALYSIS

The evaluation of the method has shown the significance of a continuous measurement of the temperatures and solar radiation, while, for practical reasons, energy measurement is performed on a daily basis and only an average air change rate for the heated space is obtainable.

As the analysis method is intended for use on a large scale and because there is no need to use complex analysis tools for identifying a pseudo-steady state model, a simple multiple linear regression model of the daily building energy balances has been adopted:

$$E_{day} = UA \cdot \int_{day} \Delta T_{i,e} - gA \cdot \int_{day} I_{sol},$$

with

$$E_{day} = Q_{heat} - C \cdot \Delta T_{day,day-1} - n \cdot V \cdot \rho_{air} \cdot Cp_{air} \cdot \int_{day} \Delta T_{i,e}$$

where $\int_{day} \Delta T_{i,e}$ and $\int_{day} I_{sol}$ are the daily integrated indoor–outdoor temperature differences and the daily integrated global solar radiation measurements; the E_{day} variable is the difference between the daily energy consumption, the possible daily variation of the energy stored by the internal thermal mass, and the daily ventilation losses (based on a test duration averaged air change rate n). Once the data set has been collected, each measurement day provides a value for E_{day}, $\int_{day} \Delta T_{i,e}$ and $\int_{day} I_{sol}$, and the multiple linear regression may be calculated, giving values for UA and gA, the equivalent global UA value and the pseudo-solar gain factor. Beyond its intrinsic simplicity, this method has the advantage of allowing a direct and reliable evaluation of the confidence intervals of the regression parameters and gives the possibility of calculating the confidence interval of E_{day} for any arbitrary chosen values of $\int_{day} \Delta T_{i,e}$ and $\int_{day} I_{sol}$.

The evaluation of the results of experiments carried out in Belgium (Belgian climate) shows that, although the total uncertainty level of the simplified method may be considered to be a bit too high, mainly because of pessimistic assumptions, the following conclusions may be drawn in case of an well-insulated dwelling:

- The test duration should be about three weeks.
- The test should preferably be undertaken during the winter, between the months of November and February, and should not be undertaken in either October or March, unless the evaluation of the impact of a slightly higher indoor constant temperature (e.g. 25°C) would lead to an acceptable uncertainty level.
- The fact that the experimental confidence intervals on the simplified consumption are rather larger than the statistical confidence intervals indicates that there is nothing to be gained from trying to develop a more complex dwelling model whereas more efforts should be invested in increasing the accuracy of the measurements.

Reference [15] gives a global summary of the method developed and of the experimental techniques applied. The next step in the development of the method has been to set up an experimental environment in an occupied house without disturbing the regular life of the occupants.

The main aim was to develop a reliable monitoring procedure, which provided the necessary input for an analysis tool. This analysis tool provides the information that makes it possible to arrive at a normalized annual heating consumption for the dwelling, based on restricted data from the heating season.

In the applied model, the overall UA value of the dwelling and the gA value are determined by means of multiple regression. For a good determination of these values, daily measurements of the ambient temperature, the mean inside temperature, the amount of heating inside the dwelling, the solar radiation and the mean ventilation rate must be available. A multiple regression calculation is used to determine the unknown parameters (UA_{global} and gA) from the model:

$$E_{day} = UA_{global} \cdot \int_{day} \Delta T_{i,e} - gA \cdot \int_{day} I_{sol}$$

with

$$E_{day} = Q_{heat} - C \cdot \Delta T_{day,day-1} - n \cdot V\rho \cdot C_p \cdot \int_{day} \Delta T_{i,e}$$

This model can only be used if it is possible to define a unique T_{int}. This is possible only when the dwelling is assumed to be one heated zone. In this case T_{int} is simply the only temperature of the heated zone.

The experimental method applied does not require detailed information on the building, but a relatively long monitoring period (at least two weeks) is required and energy readings can cause problems to, as well as disturbances resulting from, the occupancy.

In spite of its uncertainty level, the experimental method can be seen as the most reliable way to determine a normalized annual heating energy consumption since its output is relatively independent of the observed conditions and since a very limited number of assumptions have to be made regarding the building. It is, however, clear that such a method could hardly be applied on a large scale in the framework of energy certification because of the amount of work that is required.

Experimental procedure

The experimental approach adopted in the Save HELP project is a measurement-based approach (identification method).[16] The approach relies on the real behaviour of the building, which can be very different from that of its design. It also allows the influence of the climate to be distinguished from that of the occupants.

The method adopted is a time-integrated method derived from a static heat balance of the building referred to as a non-controlled method because it can be applied without any constraints on the heating schedule.

Theoretical studies have been carried out to determine which parameters should be monitored. It appears that

Table 1.6 The field experiment protocol

Variable	Recording	Time step	Destination	Caution
Energy	Reading by the occupants	Daily	Heating and appliances	Same hour each day
Indoor temperature	Wireless (e.g. Tinytag) recording sensors	≤1 hour	Representative rooms	Avoid heat sources and solar radiation
Air change rate	Perfluorocarbon tracer (PFT) measurement	Integrated	Each room	Specific protocol defined by the analyst
Solar radiation	Pyranometer	≤1 hour	Horizontal	Avoid shading
Outdoor temperature	Recording sensor (e.g. Tinytag)	≤1 hour	Outside close to the building	Protect against radiation

measuring the magnitude of the indoor air temperature, the air change rate and the power heating should lead to less than 10% uncertainty on the heating consumption, the residual error being attributed to the random behaviour of the occupant(s).

For global measurements in occupied houses the required instrumentation should provide the following data:

- temperature measurements
- ventilation measurements
- climate data measurements or data from a local weather station
- energy consumption data from measurements and/or readings.

Temperature measurements were carried out with small self-maintaining devices from Orion (Tinytag), which were able to record data at 10 minute intervals for more than a month. Five sensors were placed in the house and one device was installed outside the house to measure local ambient temperature. The measurement error in the temperature is 0.2°C.

Tracer gas measurements using PFTs were carried out in several occupied houses. The required equipment came by ordinary mail from PENTIAQ (Sweden). After the measurement period the sampling tubes and other sensors were sent back to PENTIAQ for analysis.

A description of the field experiment protocol is given in Table 1.6 and pictures of the wireless recording sensor and the PFT sources and samplers are shown in Figure 1.5.

General observations that can be made from the experiment realized:

- Occupied houses give larger uncertainties than unoccupied ones.
- Large errors were obtained for houses where large solar gains can be assumed.
- The method cannot be applied to houses heated with fuel (unless special techniques are used to measure the fuel consumption).

STEM & PSTAR*

This method has been developed in the USA. It consists of an intensive three days' monitoring of the

*Short Term Energy Monitoring (STEM) and the Primary- and Secondary-Term Analysis and Renormalization calculation procedure (PSTAR)

(a)

(b)

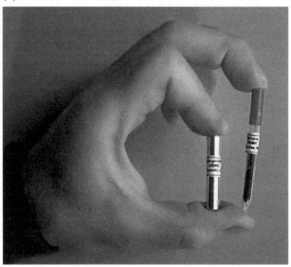

Figure 1.5 (a) Wireless recording sensor; (b) PFT sources and samplers

building. The requirements concerning the experimental technique applied are very strict. References [17–19] give a presentation of the method. The experimental techniques are described and the theoretical background and calculation methods are presented. We summarize here the main characteristics.

Theoretical background

A wide spectrum of methods can be used in thermal monitoring of buildings. The most useful methods can be broadly classified as macrostatic and macrodynamic.

Macrostatic methods are based on time integration of the energy balance of the building, with the input data (building performance and driving functions such as outdoor temperature) being similarly integrated. These methods are simple and have been used for many years in a wide variety of applications. However, these methods require long-term data, and low information content can make it difficult or impossible to reach reliable specific conclusions. Furthermore, occupancy or other schedule changes during the data period can be a major problem.

Macrodynamic methods directly employ the dynamic energy balance equation of the building, with system identification techniques used to extract some subset of the energy-balance parameters. These methods require some instrumentation in the building, such as an energy management system or a dedicated data logger. However, the requisite testing data can be acquired in a few days.

The effective leakage area is determined once at the beginning of the procedure. A data acquisition system is temporarily installed in the building. Each channel is scanned every 10 minutes and the hourly averages are stored and transferred via modem for analysis. Typical data channels are:

- six for the inside air temperature
- two for the outside temperature
- one or more for the buffer space temperature
- once each for the outside relative humidity, global horizontal solar radiation, global vertical solar radiation, total electrical power, wind velocity, furnace switch-on time and surface heat flux in case of contact with the ground.

The test protocol is quite simple. Each of the primary heat flows is forced to become dominant in the energy balance for some time period in the test. First, the steady-state conduction term is forced to be dominant by inclusion of one or more nights when the interior temperature is steady (no night setback). Second, the internal mass storage term is forced to be large by inclusion of a night of temperature decay (setback). Finally, during the test, one or more relatively sunny days must occur to ensure a good renormalization of the solar gain term.

The test procedure is programmed into the data-acquisition computer and requires four days and three nights, including set-up, take-down and analysis (Figure 1.6). The objective is to obtain steady-state conditions during the first night, to do a cool-down test on the second night and to calibrate the heating system on the third night. The tests during the second and third nights are started at midnight after a steady-state lead-in period. Daytime data are used to determine the effect of the solar gains. As far as practical, all house appliances and lights are turned off during the entire test. Up to midnight on the third night,

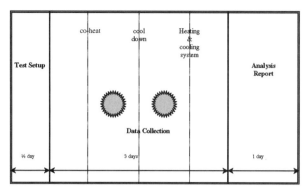

Figure 1.6 Typical three-day STEM test protocol

the furnace is off and all heat comes from several portable electric heaters individually switched on and off by the data-acquisition computer. After midnight on the third night, heat is supplied from the installed heating system, operating in response to the normal house thermostat.

For analysis, the 'renormalization factors' are introduced as simple multiplicative factors for the primary energy flows. Reference [20] describes the theory relating to the renormalization procedure. This procedure consists of defining series of 'time windows', extracting a subset of the parameters in each window sequentially with simple linear least-squares regression, and iterating to convergence. Primary- and Secondary-Term Analysis and Renormalization (PSTAR) provides a mathematical formalism for separating building energy flows, making it possible to identify the three primary thermal characteristics:

- the building loss coefficient
- the effective building mass
- the effective solar gain area.

It then uses the adjusted model to predict future building performance. In the PSTAR procedure, the heat flow into the room air is mathematically separated into several terms according to the effect causing the heat flow. The terms are as follows:

- Primary terms (to be renormalized):
  Q_1 building loss coefficient (BLC) times outside–inside temperature difference. This is the building conduction gain to the room air from the outside air under steady-state conditions. BLC is determined from the model.
  Q_2 heat flow to the room air due to a change in the inside air temperature. Calculated from the model.
  Q_3 heat flow to the room air due to solar gain. This includes the effect of solar gains through windows, heat stored in the internal mass of the building that is subsequently discharged into the room, and heat flow through the external walls due to solar absorption. Q_3 is calculated using the simulation model by setting the inside and outside temperatures equal and constant; the calculated cooling load is then the heat flow due only to solar gain.

- Primary terms (not renormalized):
  Q_4 measured heat flow to the room air due to internal gains. This is all the electrical energy into the building, including from the electric space heater.
  Q_5 heat flow to the room air due to heating of infiltration air. Calculated using the Sherman–Grimsrud model based on the leakage area; the inside–outside temperature is estimated based on the measured outside temperature and the relative humidity.
- Secondary terms (not renormalized):
  Q_6 heat flow to the room air due to a change in the outside temperature. Calculated from the model.
  Q_7 extra heat to the room air due to the lowering of the sky temperature below the outside air temperature.
  Q_8 heat flow to the room air due to conduction from an adjacent buffer space, such as a crawl space or a basement.
  Q_9 average heat flow to the ground due to direct contact with the earth.

Renormalization of the first three terms is done step-wise as follows. The energy balances descibed are averages over the time period. In each step, the previously determined values of renormalization constant are used.

1. During a period of 2 to 4 hours at the end of the night, when the inside temperature is maintained reasonably steady (called the co-heat period), Q_1, Q_4 and Q_5 are the dominant terms. That is to say the heat input from the electric heaters should approximately balance the heat losses by conduction and infiltration because heat storage, solar and other effects are small at this time. The building loss coefficient renormalization factor p_1 is determined as necessary to reconcile the observations. Specifically, Q_1 is multiplied by a renormalization constant p_1 to achieve an exact energy balance.
2. During the cool-down period, the primary heat flow into the room air results from the discharge of the building mass, because electric heat input, Q_4, is zero, or at least quite small. The mass renormalization factor p_2 is determined as required to reconcile the heat balance over the entire analysis period. Specifically, an energy balance is achieved by multiplying Q_2 by a renormalization constant p_2.
3. During the daytime hours, a major heat input is from solar gains, and electrical heat is correspondingly reduced. The solar gain renormalization factor is determined, as required, to reconcile the heat balance over the entire analysis period. Specifically, an energy balance is achieved by multiplying Q_3 by a renormalization constant p_3.

Steps 1 to 3 are repeated until the renormalization constants p_1, p_2 and p_3 stabilize. The sum of Q_1 through Q_9 (with the first three terms renormalized) is the energy imbalance Q_{net}. This should be small throughout the test period.

This method has been successfully applied over a wide range of climatic conditions. A great potential exists for use of the method for evaluating other thermal effects, such as the effects of the occupant(s), fireplace efficiency and the effectiveness of thermal storage strategies. This method can also be applied to test the efficiency of the cooling strategies realized by ventilation.

A major drawback of this method is that the experimental conditions are very strict and require that the building is unoccupied. The method also needs a short but intensive effort involving personal time.

A study describing the accuracy and the repeatability of the results obtained with this technique is presented in reference [21]. The conclusions are that, compared with other methods, this technique makes it possible to obtain quite accurate results.

Applications of the method

References [22–24] present results of experiments realized on residential buildings.

A comparison can be made between the results obtained with the tests and the interpretation of the results obtained in four houses compared with the energy consumption found by calculation methods. It is shown that important differences can be found between the two techniques and that this method can make it possible to identify the building characteristics more effectively. Some explanations are given to explain the difference between the purely calculated results and the energy consumption obtained with measured data. The influence of the solar radiation is always important in the interpretation of the results.

The STEM and PSTAR techniques can also be applied with some adaptation on commercial buildings. References [25–26] present the results of two experiences realized on this type of building. In the case of the commercial building, one of the major weaknesses of this implementation of the PSTAR approach is the single-zone nature of the modelling. This point is not developed here since the objective of this study is oriented to residential buildings.

The method has also been applied to estimating the efficiency of passive cooling strategies on residential buildings; the results of a case study are given in reference [27].

NEURAL NETWORKS
Background theory

The application of neural network techniques is very wide. In the scope of this study, neural network techniques are used to predict the future thermal behaviour of a building when sufficient data from the past are known. References [28–30] give short but comprehensive descriptions of the theory relating to the neural networks, we summarize here the main characteristics of the technique.

Conceptually, a neural network (NN) is made up of interconnected nodes arranged in at least three layers (Figure 1.7).

The input layers merely receive the data patterns related to the input variables. The number of input nodes

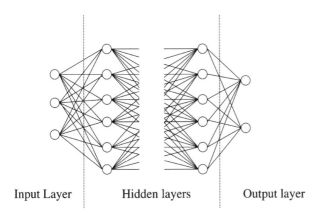

Input Layer | Hidden layers | Output layer

Figure 1.7 Neural network scheme

consequently equals the number of measured data values. The hidden and output layers both actively process data. To do that, each neurone acts as a connection. Associated with each connection is an adjustable value called a weight. Basically, a node calculates the weighted sum of its input, then passes the sum through a function to produce a result. This transfer function is typically a steadily increasing S-shaped curve. The attenuation at the upper and the lower limits of the S constrains the sum within fixed limits. The back-propagation algorithm is able to change the values of its weight in response to errors. It then compares the desired and the actual results.

The differences are the errors in the output layers, which the network passes back to the hidden layer using the same weighted corrections. After each output node and each hidden node finds its error value it adjusts the weight to reduce its error. The equation that changes the weights – called the delta rule – is designed to minimize the network's sum-squared error. After training, the network should be tested with known data that has not been used during training. The accuracy of the network with a pattern outside the training set is called generalization and indicates its reliability in an application. After training and testing, the network is ready to process unknown data. Basically, applying a pattern to the input produces a corresponding pattern at the output. The network therefore acts as a model of a function, mapping input patterns to output patterns. It learns this association solely from the training data, even if the equation describing the function is non-linear, unknown or both.

In the summer 1993, an energy modelling competition was organized by ASHRAE, called 'The great energy predictor shoot-out'.[31] The purpose was to evaluate different methods for predicting hourly energy use. The competition was supplied with four months of historical data and was asked to predict the following two months. Five of the six most successful contenders used different neural network techniques.

Applications of the neural network technique

Neural networks are used in some cases to determine experimentally the annual space heating demand.

References [28] and [32–33] present the results of case studies. The method applied uses NN techniques to obtain models that are capable of predicting the supplied heating demand of small family buildings. The models are based on a limited amount of measured diurnal performance data for a short time period. The results obtained show that, with access to only measured data of supplied space heating demand and climatic data in terms of indoor and outdoor temperatures, the supplied space heating demand can be predicted to within 5–10% on an annual basis.

The method has been applied on 10 single-family houses situated 700 km north of Stockholm, Sweden. Limited monitoring has been realized. The data coming from the monitoring were meteorological data and building performance data; energy demand, outdoor and indoor temperatures were measured at the buildings. Other meteorological data, such as wind, relative humidity, air pressure and solar irradiation, were measured at a nearby airport.

This method is a combination of a neural network and a quasi-physical description, which requires only access to the average daily outdoor and indoor temperatures and to the space heating demand for a limited period of time. Despite the diversity of the different buildings in term of building design, ventilation system, control system and local climate, this approach yielded a rather robust performance. Except for periods when the supplied space heating demand was very small, the approach yielded fair predictions. *The method requires only performance data on a level that can be supplied by most house owners.* The method can also be used for both the domestic and the total energy demand of a building.

The traditional method for simulating the energy demand in buildings is based on implementations of building characteristics in more or less ideal physical sub-processes. Although this method gives a good physical understanding, it requires a large number of parameters and different degrees of idealization and simplification. Other common modelling techniques based on measured performance data are, for instance, time-series analysis and statistical methods. These methods require no detailed knowledge of the building, but, instead, the results can be difficult to interpret in physical terms.

An advantage of neural network techniques is their good ability to map non-linear dependencies between input and output data and, also, that this can be done without any conceptions of intrinsic relations in the presented data. A drawback of neural networks, as well as many other non-analytic models, is the difficulty of interpreting the model in practical terms, as well as the limitations in the accuracy of the predictions for events outside the data sets.

Significant for neural networks is that they do not demand any explicit rules or knowledge for processing data. For neural networks, the rules are built within the system.

With this technique the supplied space heating demand can be predicted with an accuracy of about 5%. *Although this technique does not need heavy monitoring and the results have a fair accuracy, the exploitation of the data requires experience and feeling and is relatively time-intensive.* The model needs first to be trained with the available data before it is used with the meteorological conditions of the reference year. *This method could be difficult to apply for certification purposes.*

Reference [34] presents a similar approach. In this model, the behaviour of the inhabitants in terms of the domestic load is included. In order to do this, the predicted annual variation of the domestic load based on measured performance data, together with an assumption of an annual profile of the domestic load obtained from a more extensive measurement, is used. In addition to predicting the supplied space heating demand, the method predicts the total energy demand. As model input, the model uses the difference between indoor and outdoor temperatures, a building climate perspective obtained from dynamic energy simulation software and a measure of inhabitant influence in terms of a predicted domestic load. The predictions of the domestic load were based on measurements performed in a separate single-family building. Compared to a previous model, based mainly on a building-climate perspective, the deviation between the predicted and measured annual variation decreased from 7.5% to 4%.

Neural networks can also be used to analyse the interactions between the heating system and the energy load[35] and to analyse the interactions between heating and domestic load in an occupied single-family building.[36]

Supplementary examples of applications of the neural network technique

Neural networks can also be used in other types of application.

Reference [29] presents a case study of an application of the neural network technique to predicting the building energy consumption of a building in the future without knowledge of the immediately past energy consumption. Such a prediction is of value when estimates of what a building, retrofitted with energy conservation features, would have consumed had it not been retrofitted.

Reference [37] presents a supplementary application of neural networks. It is used here to produce a utility consumption prediction indicator for a commercial building. The predictors are afterwards used in expert systems to realize diagnostics for Heating, Ventilation and Air Conditioning (HVAC) systems. The procedure used to find the predictors uses historical data, such as weather data, building occupancy and other factors known to affect energy consumption.

Reference [30] presents an application of neural networks in the case of the control of the heat power delivered to a building. Four parameters have to be introduced as input to the NN: outdoor air temperature, solar radiation, indoor air temperature and energy consumption at time $t-1$. The result of the NN is the energy consumption at time t, which is the heat power. The NN technique delivers accurate results and could be developed for other control purposes.

The last application of neural networks presented here concerns the use of the CO_2 concentration to predict energy consumption.[38]

OTHER RELATED TOPICS

In this section we summarize some other interesting articles.

Miscellaneous

Reference [39] describes the results of a set of simulations of the total energy consumption of two buildings by taking into account the uncertainties of the different parameters playing a role in the energy consumption (climate, building and inhabitants). These simulations and the statistical utilization of them make it possible to assess the influence of these parameters on the total variability of the energy consumption. Some interesting results are that, without knowing the inhabitant behaviour, it is impossible to predict the total energy consumption more accurately than ±15–20%. Furthermore, the heating and ventilation energy has an uncertainty of ±25–40% if the inhabitant's behaviour is unknown. The colder the climate, the better the accuracy of the heating consumption.

This article shows that certification of the energy consumption by a descriptive method will unavoidably be affected by the uncertainties in the different parameters (climate, building and inhabitants).

Reference [40] describes the origin of the error encountered when energy simulations are realized with energy signature models. It appears that including the solar radiation in the model can significantly improve the accuracy of the results obtained.

The article shows that the calculation and the introduction of the weather correlation of solar radiation I and outside temperature T in the simplest model for every frequency of data confirms quantitatively that omission of the solar radiation is the main cause of error in the estimate of the parameters of the building.

Conventional building load simulation models give performance predictions based upon physical properties of materials and specifications of equipment, the real values of which are frequently uncertain. Interactions between the building and the inhabitants or the environment are also imperfectly known. In another approach, some models use measured data; these models, sometimes called measurement-based energy models, extrapolate building parameter estimations and energy performance predictions over an extended period. The available data may only consist of energy consumption and weather characteristics.

Experimental techniques

There are some other references to experiments realized using monitoring techniques (see references [41–47]). The purposes of these tests can be very different and some articles are interesting but less relevant to this study. We do not summarize these articles here.

Calculation methods and parameter estimation

We give here some other references to calculation methods and techniques for estimating the parameters used in the calculation methods. Since this is not the main topic of this report, we do not give much detail.

Reference [48] describes an experiment carried out to test and validate a measurement-based model with 'synthetic' consumption data generated by a conventional building load simulation model. The model tested is SUBMET. This technique makes it possible to test the algorithm, to define what is the best form to give to the data to obtain the best accuracy, to define the best accuracy that can be expected from the model, etc. Note that this technique could also be used to test measurement-based models for cooling energy use.

The article begins by giving a description of different existing measurement-based building energy models. Measurement-based models typically derive parameters of energy performance from the measured data. These parameters are combined with weather data and other physical characteristics of the building to predict thermal performance or average annual space heating consumption. There are both dynamic and static measurement-based models. Among the dynamic models is BEVA (Building Energy Vector Analysis). Static models PRISM and SUBMET aim to predict the average space heating use in new, low-energy single-family homes. SUBMET is a regression-based model but requires more detailed information than PRISM.

Reference [49] presents a comparison of different methods for dynamic analysis of measured energy use. The theory and the interpretation of the different methods are presented. The methods are thermal networks, the ARMA (AutoRegressive Moving Average) model, differential equations and modal analysis.

The same example has been tested with the four models. The difficulties related to the four models are presented. This article gives many bibliographic references to publications describing each method separately. The article deals more with calculation methods than with the experimental techniques used to collect the data.

Further information can be found in references [50–59].

CONCLUSIONS

There are several experimental methods that could be useful to the procedure currently developed. However, most of the references that we have found were more focused on the calculation methods rather than on the detail of the experimental monitoring used.

Because of the various advantages offered by the methodology used, the technology of the Energy Barometer project seems to be the most appropriate for the method currently developed. The use of the Internet and a modem to collect the data is the most promising technology available for developing an accurate method with reasonable exploitation costs.

The measuring techniques employed in the Save HELP project is a light monitoring approach that could easily be employed in other circumstances. The only sensitive point here is the measurement of the ventilation rate. The technique used (PFT) requires care, needing qualified personnel with experience and a sensitive approach.

The procedure developed in the STEM and PSTAR method is very detailed and much relevant information can be found in the literature. The protocol used is very strict and the houses have to be unoccupied in order to make measurement convenient. The results obtained are probably the most accurate of the different methods described. This type of prescriptive measurement protocol can probably not be adopted in the developed methodology because of the complexity of the set-up and of the calculation phase.

The experimental techniques used in the neural network methods have essentially been developed in order to provide data for the calculation methods. However, it should be noted that the type of networks adopted can influence the results. With these types of technique, the choice of the data used to train the network can also have an influence on the results obtained. For each case it is necessary to adapt/train the model used for that specific case. It should also be noted that the results obtained can be difficult to interpret in a physical sense. These types of technique require highly qualified personnel and are probably still best kept for research purposes rather than for certification methods.

Some other methods and related topics have been identified and are included in the references.

We have not been able to find articles describing methods used to characterize the functioning of active cooling. Some of the calculation techniques described could be extended for this purpose but no experiment tests have been described.

Part 2. In situ evaluation of UA and gA values – an overview of possibilities and difficulties

INTRODUCTION

This part of the report gives an overview of the state of the art in relation to the on-site evaluation of the UA and gA values of buildings.

DIRECT MEASUREMENTS OF U-VALUE

It is possible to measure with quite good accuracy (5–10% or better) the U value of building components. Such measurement is typically done by using a heat flow meter installed on the inner surface of the component and temperature sensors on both surfaces.

The procedure for determining the 'surface-to-surface thermal resistance' is described in the European standard prEN 12494.[60]

It should be noted that, in the case of transparent components (glazing, etc.), it is not easy to carry out a measurement if there is incident shortwave radiation, because of problems of accuracy of the measurements realized (e.g. glazing surface temperature).

Measurements in climate chambers

HOT BOX APPARATUS

The hot box measurement technique using a guarded or calibrated hot box (EN ISO 8990/EN 1946-4)[61] is well known for identifying the UA value of building components.

OUTDOOR TEST FACILITIES

Since the 1980s, outdoor climate chambers have received increased interest:

- In the USA, the MOWITT unit developed at Lawrence Berkeley Laboratory (LBL) has to be mentioned.
- In Europe, efforts have been concentrated in the framework of the European PASSYS, PASLINK (Figure 1.8), COMPASS, PV-HYBRID-PAS and IQ-TEST projects.

Detailed monitoring campaigns in non-occupied buildings

GENERAL

The verification of thermal performances in real buildings seems very attractive. In the framework of the PLEIADE project,[62] one of the aims was to verify in situ the reliability of the calculation procedures for estimating the global transmission losses at building level, calculated according to the Belgian standards NBN B62-301, NBN B62-002 and NBN B62-003.[63,64,65]

As far as the authors are aware, the procedure was probably unique at the time it was first applied in the PLEIADE dwelling, when specific attention was given to the minimization of the uncertainties. Basically, it consists of four essential features:

1. *A refined procedure for monitoring climate parameters and energy consumption data.* The monitoring measures not only all room temperatures, the outdoor temperature, the solar radiation and energy use, but also the air flow rates with multiple tracer gas measurements and the water vapour balance of the house.
2. *Physical measures for minimizing solar gains during the monitoring.* Based on the error analysis in simulated experiments, evidence was provided that carrying out experiments involving important solar gains (as is the case in PLEIADE) in combination with low transmission losses would lead to a very large uncertainty in the predicted thermal transmission losses for the whole building. This uncertainty can be significantly reduced by minimizing the solar gains. In practice, this was done by having self-adhesive white sheets on the outside of the glazing and by putting up a scaffolding, covered by an opaque canvas, around the façades (see Figure 1.9).
3. *Refined control of the indoor climate and of certain boundary conditions.* The approach used in NBN B62-301 for determining the transmission losses at building level gives different weighting factors for the losses through the components in direct contact with outside (weighting factor WF = 1), through the floors in contact with the basement (WF = 2/3) and through the common walls (WF = 0). A simple co-heating experiment does not allow a distinction to be made between the three types of heating losses measured

Figure 1.8 A picture of a PASLINK test cell

Figure 1.9 Specific measures in the PLEIADE building for increasing monitoring accuracy. *Source:* BBRI[66]

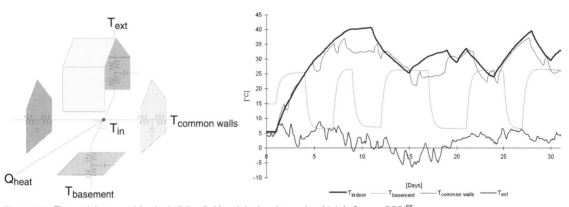

Figure 1.10 The model assumed for the building (left) and the heating regime (right). *Source:* BBRI[66]

during the experiment. In order to allow a correct weighting, a methodology was developed to have some control of the temperature profiles in the basement (T_{basement}) and of the outside surface temperature of the common wall ($T_{\text{common walls}}$) of the PLEIADE dwelling. The latter was achieved by integrating electrical heating foils into the construction of the PLEIADE dwelling (Figure 1.9). As is illustrated in Figure 1.10, very different temperature profiles were used so that there was a low correlation factor between the various temperature profiles. Moreover, high internal temperatures were applied to increase the temperature difference between inside and outside.

4. *The use of system identification techniques for data analysis.* Based on the methodologies developed in the PASLINK projects, system identification techniques are applied to the measured data.

PRACTICAL RESULTS

For the monitoring period (January 1995), the relative energy use to compensate for the transmission losses, the ventilation losses and the drying out of the dwelling (the monitoring period was less than one year after construction) is given in Figure 1.11. Only 71% of the total identified

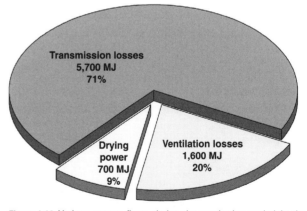

Figure 1.11 Various energy flows during the monitoring period in the PLEIADE dwelling. *Source:* BBRI[66]

energy use served to compensate for the transmission losses. The derived thermal insulation level (according to NBN B62-301[63]) was K27* (95% confidence interval: ±4), which agrees very well with the theoretical derived value of K28 (Figure 1.12). If the monitoring had been limited to the

*The K-level represents the thermal insulation level of a building. For buildings with a low capacity, it represents 100 times the average U value of the building $K = 100 \times$ Um for buildings with V/A $\leqslant$ 1 M

transmission losses, and no ventilation losses or energy use for drying out of the building assumed (Figure 1.11), a thermal insulation level K38 would have been calculated, a value 40% higher than the calculated one.

The probable conclusion would then have been that the theoretical calculations strongly overestimated the real performance.

It is important to draw attention to the treatment of the 'common walls' of the PLEIADE dwelling. In principle, the thermal insulation level neglects the heat losses through the common walls. The PLEIADE dwelling was built before the adjacent dwellings were constructed.

According to NBN B62-301, all walls on the border line with the neighbours have to be considered as a common wall. In reality (and allowed by the urban regulations), one of the neighbours has used another building form (Figure 1.13), as a result of which a part of the assumed common wall becomes an external wall (about 25 m^2). It is clear that such a modification can significantly increase the transmission losses (especially if this wall is not insulated).

The authors believes that the experiences in the PLEIADE dwelling as well as in other buildings and in the PASLINK test environment make it possible to draw the following conclusions with respect to empirical model validation of thermal simulation programmes:

- Rather good agreement between prediction and monitoring results can probably be found in many cases if a number of conditions are met: the overall transmission losses have to be calculated based on the real situation and take into account the precise location of adjacent buildings; the insulation has to be correctly installed and the assumed λ values should be based on the measured values for the materials applied in the building.
- Important differences may occur if certain practical aspects of the monitoring are neglected, e.g. the building form of neighbouring buildings (see Figure 1.13), thermal bridges, etc.

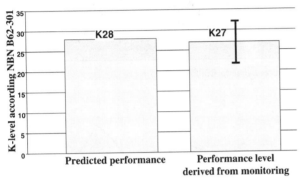

Figure 1.12 PLEIADE: Comparison between predicted and measured thermal insulation levels. *Source:* BBRI[66]

Figure 1.13 A 'common' wall in PLEIADE (Architect: P. Jaspad)

In the literature, (rather) important differences between theoretical calculations and monitoring results are sometimes reported. This would have also been the case for PLEIADE if the ventilation losses, the drying out of the building, the losses through the common walls, etc. had not been correctly monitored. The authors suggest that one must be extremely careful with statements concluding that monitoring results have 'proven' that certain simulation procedures are unreliable. Discrepancies are probably quite often more related to the quality of the monitoring activities than to the quality of the simulation tool.

Detailed monitoring in occupied dwellings

BRIEF OVERVIEW OF INTERESTING STUDIES

Save HELP

The Save HELP project has already been discussed earlier in this chapter.[14] Here we focus more on the results of the calculation method.

This European study describes a simplified method for the energy certification of unoccupied buildings. First, the applicability of the proposed method was assessed by means of numerical simulations, making it possible to define the experimental requirements. Second, the sensitivity study focused on the seasonal, duration and building type dependency of the method accuracy. Third, the resulting method was applied to a well-known unoccupied test dwelling, making it possible to check in practice the relevance and validity of the method and of its design.

The basic idea of the method is to derive a simplified thermal model of the dwelling to be certified by means of a limited monitoring of the building (energy consumption, indoor and outdoor climate) with uncontrolled indoor conditions (constant temperature strategy). This simplified model is then used to calculate, on the basis of typical climatic data and a reference occupancy effect, a standardized heating energy consumption for the heating season. As the objective of the Save HELP project is to design a certification procedure and to assess its relevance in a straightforward way, several simplifying assumptions were made. Among them, an electrically heated dwelling with a heating system having an efficiency of 100% was used as the base case. Although the method proposed indicates how to deal with lower efficiencies, its applicability remains limited to buildings equipped with a heating system that allows a daily reading of the heating energy consumption.

PhD thesis by Zoltan Somogyi

An interesting approach has been developed in the PhD thesis of Z. Somogyi,[67] which takes a similar approach to that of the Save HELP project.

This approach takes into account (Figure 1.14):

- transmission and ventilation losses
- solar gains
- estimation of internal gains.

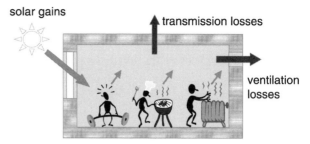

Figure 1.14 Energy flows involved in the identification process

Methodology

The essential features of the methodology are:

- detailed continuous measurement of indoor and outdoor temperature
- on-site measurement of solar radiation
- estimation of the shading impact of the environment for the various windows
- estimation of the ventilation rate (by the passive tracer gas method, by the pressurization method in combination with the LBL-model, etc.)
- daily recording of energy consumption
- estimation of the thermal capacity of the dwelling.

These measurements are typically made for 20–40 days. Based on daily average values for all parameters, system identification is carried out, so that RC models such as the one presented in Figure 1.15 are identified. The main result of a single identification run is an estimation of the UA and gA values together with their confidence intervals.

As part of the whole concept, correct estimation of the confidence intervals is very important. A Monte-Carlo approach is therefore applied, whereby for each input variable a probability function is estimated. A hundred different data sets are generated (Figure 1.16) so that there are 100 sets of UA and gA values with their respective confidence intervals.

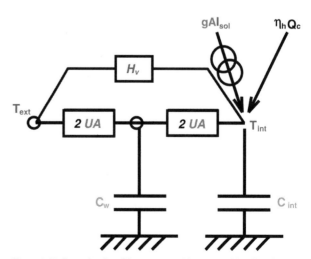

Figure 1.15 Example of an RC system used for system identification

In practice, one observes:

• large confidence intervals
• a high correlation between the different sets of UA and gA values.

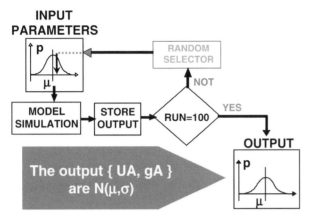

Figure 1.16 The concept of the Monte-Carlo approach for error analysis

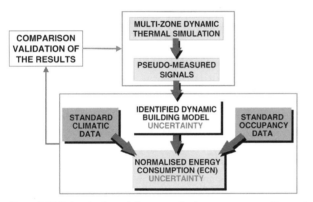

Figure 1.17 Determination of the normalized energy consumption and related confidence intervals

The high correlation clearly indicates that it is not necessary to make accurate estimations of the individual values of UA and gA. A so-called normalized energy consumption is therefore determined, which estimates the annual energy consumption for a standard year and for fixed indoor conditions (Figure 1.17).

Error analysis

In Figure 1.18, the results are shown for an unoccupied dwelling in the Belgian climate. The large differences in accuracy can clearly be seen.

In Figure 1.19, the impact of the average indoor temperature is shown. The higher the average indoor temperature, the better the accuracy.

On site, the method was applied in one unoccupied dwelling (Figure 1.20) and in two occupied dwellings (Figure 1.21).

The results and confidence intervals for the normalized energy consumption are shown in Figure 1.22.

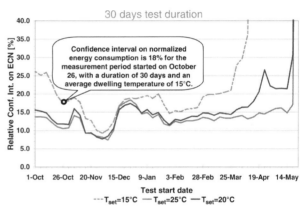

Figure 1.19 The impact of average indoor temperature on accuracy

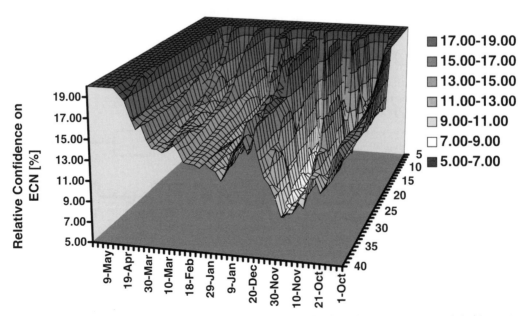

Figure 1.18 The accuracy of the identified normalized energy consumption for various measurement periods (the y-axis shows the duration of the measurement in days)

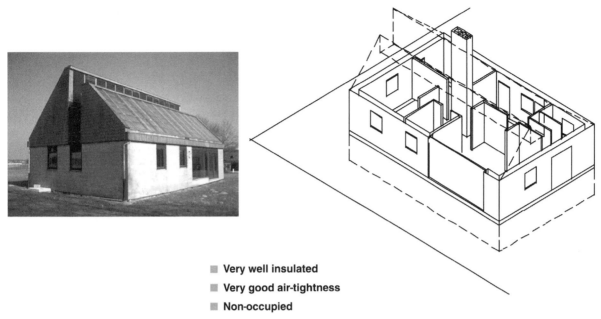

■ **Very well insulated**
■ **Very good air-tightness**
■ **Non-occupied**

Figure 1.20 The unoccupied dwelling used for monitoring

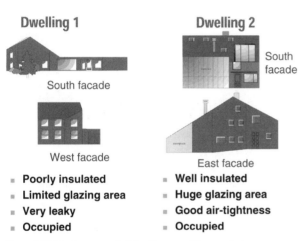

Dwelling 1

South facade

West facade

■ **Poorly insulated**
■ **Limited glazing area**
■ **Very leaky**
■ **Occupied**

Dwelling 2

South facade

East facade

■ **Well insulated**
■ **Huge glazing area**
■ **Good air-tightness**
■ **Occupied**

Figure 1.21 The two occupied dwellings used for monitoring

The following conclusions are important:

• The best results are found in the non-occupied dwelling (IDEE). This is clearly due to the relatively small uncertainty in the internal gains, the ventilation rates and the average building temperature.
• The relative accuracy is better in the poorly insulated dwelling (ECSII). This is also quite logical since the losses are relatively large, which allows a more accurate estimation of the normalized energy consumption.

The impact of climatic region

NORMALIZED ENERGY CONSUMPTION

The authors are not aware of any systematic studies relating to the impact of the climate (e.g. warm region in Southern Europe, moderate region in Western Europe or cold region in Northern Europe) on the relative accuracy of the normalized energy consumption.

However, it is almost certain that there are major differences in the relative accuracy:

• In the north of Europe, in winter time there are high transmission losses (and large temperature differences) and very low solar gains. As a result, the difference between losses and gains can be estimated with a good accuracy.
• In southern warm regions, in winter time the absolute values of the losses are relatively low (and temperature differences are relatively small). Moreover, there are quite often rather substantial solar gains. As a result, the difference between losses and gains cannot be estimated with the same relative accuracy as in the first case.
• In moderate climates, the situation is probably somewhere in between the previous two cases.

The authors therefore believe that the best relative accuracy for the normalized energy consumption can be obtained in cold climates.

INDIVIDUAL VALUES FOR LOSSES AND GAINS

Northern cold climates

In line with the observations given above, good relative accuracy is expected for the loss value (transmission + ventilation) in northern cold climates if measurements are carried out during the cold winter months. Indeed, the losses are almost equal to the heating demand. However, relatively poor accuracy is probably found for the gA value because of the low value of the solar gains, both in absolute terms and in comparison with the losses.

Southern warm climates

The measured energy consumption is often the difference between two values (losses and gains) of rather similar

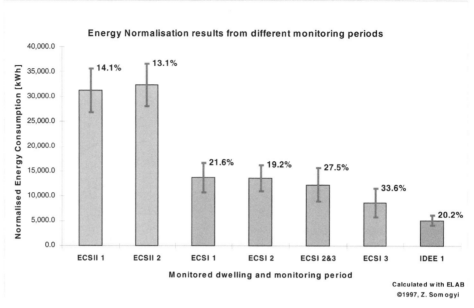

Figure 1.22 Results for one unoccupied dwelling and two occupied dwellings

Figure 1.23 Temperature datalogger

Figure 1.24 The ESTI sensor

magnitude. It is therefore not possible to estimate the loss and gain factors with a reasonable accuracy.

Monitoring equipment

TEMPERATURE

Long-term temperature measurements were in the past only possible with rather heavy and expensive data loggers in combination with temperature sensors. During the last decade, however, small data loggers have become available and very affordable prices are now combined with a large storage capacity and quite good accuracy. One can now buy, for about €100 or less, data loggers that allow the storage of up to 100,000 measurement points. Two typical data loggers are shown in Figures 1.5 and 1.23. The data can subsequently be read by a normal PC.

SOLAR RADIATION DATA

In the past, on-site measurement of solar radiation data was quite expensive. Here also, there have been some interesting developments. In particular, the ESTI sensor (Figure 1.24), developed by JRC Ispra, in combination with the data logger technology mentioned in the previous section makes on-site measurements rather affordable.

AIR CHANGE RATE

The air change rate can be measured with active or passive tracer gas techniques. In occupied buildings, active tracer gas measurements are not possible. In such cases, passive tracer gas measurements (Figure 1.5) can be an alternative. On average, the accuracy is of the order of 20–30%.

ENERGY USE

The measurement of energy use can be difficult in some cases. A summary of the different adopted techniques available to measure of the energy consumed is given by Akander, Johannesson *et al.*[68]

Electricity and gas

In principle, daily recording is no problem. In practice, monitoring equipment has to be installed and security aspects have to be taken into account.

Fuel, wood, etc.

It is often not possible to monitor these.

CONCLUSIONS

1. Direct on-site measurement of U value is possible. A European standard (prEN 12494) is available and describes the required procedure.[60]
2. *Reliable* identification of the UA and gA values in unoccupied buildings is not possible, because of the often high correlation between the identified UA and gA values. As a result, the confidence intervals on both parameters are often (very) large. Moreover, it is often not possible to estimate the ventilation losses with a reasonable accuracy.
3. Instead of identifying the individual UA and gA values, it is easier to identify the normalized energy consumption.
4. In the case of occupied buildings, there are several additional problems (more important internal gains, which are often not well known, the influence of user patterns, the opening of doors and windows, etc.), because of which the uncertainty in the identified normalized energy consumption substantially increases. In occupied buildings, it does not seem at all possible to identify the UA and gA values with a reasonable accuracy if such identification is to be based on the measured energy use by the building itself.
5. The following trends are important:
   - The relative accuracy is typically worse in better-insulated buildings.
   - The relative accuracy is typically worse in southern warm climates.
   - The relative accuracy is typically worse for short measurement durations.

REFERENCES
The university projects & the Energy Barometer

1. Bostadsdepartementet, 1980, *Energispareffekter i Bostadshus där Åtgärder Genomförts med Statligt Energisparstöd.* Expertbilaga 5 till SOU 1980:43 – Program för energihushållning i befintlig bebyggelse. Ds Bo 1980:8.
2. Elmroth A, Hjalmarsson C, Norlén U, Rolén C, *et al.*, 1989, *Effekter av energisparåtgärder i bostadshus.* Rapport R107:1989, Byggforskningsrådet, Stockholm, Sweden.
3. Westergren K-E, Högberg H and Norlén U, 1998, 'Monitoring energy consumption in single family houses', in *Energy and Buildings*, 29, 247–257.
4. Westergren K-E, 2000, *Estimation of energy need for heating in single-family houses*, R&D-report No. 3, R&D committee, Royal Institute of Technology, Stockholm.
5. Norlen U, 1985, 'Monitoring energy consumption in the Swedish building stock', *Proceedings of Conference on Optimisation of Heating Consumption, Prague*, Swedish Institute for Building Research, Gävle.
6. Hammarsten S, 1987, 'A critical appraisal of energy signature models', in *Applied Energy*, 26(2), pp. 97–110.
7. Hammarsten S, 1984, 'Estimation of energy balances for houses', *Bulletin* M84:18; Doctoral thesis, The National Swedish Institute for Building Research, Gävle.
8. Westergren K-E, 2000, Energy use for heating in houses with a heat pump; Working paper series No 47 Research and Development Committee at University of Gävle in cooperation with The Royal Institute of Technology, Stockholm.
9. Westergren K-E and Waller T, 1998, 'Virtual Housing Laboratory, A system for simulating the energy use for heating in single family houses', Working Paper No. 1, R&D Committee, Royal Institute of Technology and University of Gävle, Sweden.
10. Boman C A, Jonsson B-M and Mansson L-G, 1993, *Eleffektiva smahus. Värme och vitvaror-fullskaleförsök*, Rapport TN:42, The National Swedish Institute for Building Research, Gävle.
11. Lindfors A, Westergren K-E and Lilliestrale M C J, 1998, 'An Internet based system suitable for European wide monitoring energy use', in *EPIC 98, Lyon, France (19–21 November 1998)*, Vol. 2, pp. 606–611.
12. http://www.ericsson.com
13. Westergren K-E, Högberg H and Norlén U, 1996 (revised 1997), *An Energy Barometer for Sweden. Pilot Study and Proposal*, Build Environment, Royal Institute of Technology, Gävle, Sweden.

Save HELP

14. Martin S, Wouters P and L'Heureux D, 1996, *Evaluation of a simplified method for the energy certification of non-occupied buildings, Final Report*, Save HELP Project, Belgian Building Research Institute, Brussels.
15. Bloem J J and Martin S, 1998, 'A pseudo dynamic analysis tool for thermal certification of dwellings', in *EPIC 98, Lyon, France (19–21 November 1998)*. Lyon: Ecole Nationale des Travaux Publics de l'Etat, 1998, Vol.2, pp. 403–408.
16. Richalet V, Neirac F P, Tellez F, Marco J and Bloem J J, 1998, 'HELP (House Energy Labelling Procedure): Methodology and present results', in *EPIC 98, Lyon, France (19–21 November 1998)*. Lyon: Ecole Nationale des Travaux Publics de l'Etat, 1998, Vol.2, pp. 129–134.

STEM and PSTAR

17. Balcomb J, Burch J and Subbarao K, 1992, *Short-Term Energy Monitoring: An Overview*, Golden, CO, Solar Energy Research Institute.
18. Subbarao K, 1988, *PSTAR – Primary and secondary terms analysis and renormalization: A unified approach to building energy simulations and short-term monitoring*, Report SERI/TR-254-3175, Golden, CO, Solar Energy Research Institute.
19. Subbarao K and Burch J, 1985, *Macrodynamic Theory and Short-Term Measurements for the Thermal Assessment of Buildings*, Report SERI/TR-253-2854, Golden, CO, Solar Energy Research Institute.
20. Subbarao K, 1984, BEVA *(Building Element Vector Analysis) – A new Hour-by-Hour Building Energy Simulation with System Parameters as*

Inputs, Report SERI/TR-254-2195, Golden, CO, Solar Energy Research Institute.

21. Burch J D, Subbarao K, Hancock C E and Balcomb J D, 1989, *Repeatability and predictive accuracy of the PSTAR short-term building monitoring method*, Report SERI/TR 254-3606, Golden, CO, Solar Energy Research Institute.

22. Balcomb J D, Burch J D and Subbarao K, 1993, 'Short-term energy monitoring of residences', in *ASHRAE Transactions*, 99(2) 935–946.

23. Subbarao K, Balcomb J D, Burch J D, Hancock C E and Lekov A, 1990, 'Short-term energy monitoring summary of results from four houses', in *ASHRAE Transactions*, 96(1), 1478–1483.

24. Subbarao K, Burch J, Hancock C E and Lekov A, 1988, *Short Term Energy Monitoring (STEM): application of the PSTAR method to a residence in Fredericksburg, Virginia*, Report SERI/TR3356, Golden, CO, Solar Energy Research Institute.

25. Burch J, Subbarao K, Lekov A, Warren M and Norford L, 'Short-term energy monitoring in a large commercial building', in *ASHRAE Transactions*, 90, 1459–1477.

26. Balcomb J D, Burch J D, Subbarao K and Hancock C, 1994, 'Short-term energy monitoring for commercial buildings', in *Proceedings of ACEEE Summer Study on Energy Efficiency in Buildings*, Vol. 5, *Commissioning, Operation and Maintenance*. Washington: American Council for an Energy-Efficient Economy, 1994.

27. Byars N, Hancock C, Anderson R and Balcomb J., 1990, 'Experimental evaluation of passive cooling using PSTAR', in *15th National Passive Solar Conference, American Solar Energy Society, 19–22 March 1990, Austin, Texas*. American Solar Energy Society, 1990.

Neural networks

28. Olofsson T, Andersson S and Östin R, 1998, 'A method for predicting the annual building heating demand based on limited performance data', in *Energy & Buildings*, 28, 101–108.

29. Kreider J F, Claridge D E, Curtiss P, Dodier R, Haberl J S and Krati M, 1995, 'Building energy use prediction and system identification using recurrent neural networks', in *Journal of Solar Energy Engineering*, 117, 161–166.

30. Cammarata G, Fichera A, Marletta L, Micali A and Vagliasindi C, 1998, 'Control of indoor air temperature in buildings based on neural networks', in *EPIC 98, Lyon, France (19–21 November 1998)*. Lyon: Ecole Nationale des Travaux Publics de l'Etat, 1998, Vol.2, pp. 630–634.

31. Kreider J F and Haberl J S, 1994, 'Predicting hourly building energy use: the great energy predictor shoot-out – overview and discussion of results', in *ASHRAE Transactions*, 100, 1104–1118.

32. Andersson S, Olofsson T and Östin R, 1996, 'Predictions of energy demand in building using neural network techniques on performance data', in *Proceedings of the 4th symposium on Building Physics in the Nordic countries, Espoo, Finland*, Vol. 1, 51–58.

33. Olofsson T and Andersson S, 2000, 'Long-term energy demand predictions based on short-term measured data', in *Energy and Buildings*, 33 (2), pp. 85–91, 2002.

34. Olofsson T, Andersson S and Östin R, 1998, 'Energy load predictions for buildings based on a total demand perspective', in *Energy and Buildings*, 28(1), 109–116.

35. Olofsson T and Andersson S, 1999, 'Analysis of the interaction between heating and domestic load in occupied single-family buildings', in *Proceedings of the 5th symposium on Building Physics in the Nordic Countries, in Gothenburg, Sweden*, Vol. II, 473–480.

36. Olofsson T and Andersson S, 2000, 'Overall heat loss coefficient and domestic energy gain factor for single-family buildings', in *Energy and Buildings*, 37 (11), pp. 1019–1026, 2002.

37. Kreider J F and Wang X A, 1991, 'Artificial neural network demonstration for automated generation of energy use predictors for commercial buildings', in *ASHRAE Transactions*, 97(2), 775–779.

38. Olofsson T, Andersson S and Östin R, 1998, 'Using CO_2 concentrations to predict energy consumption in homes', in *Proceedings of the 1998 ACEEE summer study of energy efficiency in Buildings. American Council for Energy-Efficient Economy, Washington DC*. Washington: American Council for an Energy-Efficient Economy, 1994, Vol. 1, 1211–1222.

Other related topics

39. Pettersen T D, 'Variation of energy consumption due to climate, building and inhabitants', in *Energy and Buildings*, 21, 209–218.

40. Flouquet F, 1992, 'Local weather correlations and bias in building parameter estimates from energy signature models', in *Energy and Buildings*, 19, 113–123.

41. Stephenson D G and Millas G P, 1967, 'Cooling load calculations by thermal response factor method', in *ASHRAE Transactions*, 73, pp. 311–317.

Other measurement techniques and monitoring experiences

42. Burch J D, 1986, 'Building thermal monitoring methods', in *Passive Solar Journal*, 3(2), 149–177.

43. Mazzuchi R P, 1992, 'End-use profile development from whole-building data combined with intensive short-term monitoring', in *ASHRAE Transactions*, 98(1), 1180–1186.

44. Palmiter L, Toney M and Brown I, 1988, *Preliminary evaluation of two short-term building test method*, Ecotope Inc. Seattle.

45. Subbarao K *et al.*, 1985, 'Measurement of effective thermal capacitance in buildings', in *Proceedings ASHRAE/DOE/BTECC Conference on Thermal Performance of the Exterior Envelope of Buildings, Clearwater Beach, FL, December 1985*. Atlanta: American Society of Heating, Refrigerating and Air-conditioning Engineers.

46. Flouquet F and Richalet V, 1989, 'Analyse comparative de méthodes pour l'évaluation des performances énergétiques des bâtiments sur site', in *Proceeding Conference on Science and Technology at the Service of Architecture, Paris*. Dordrecht: Kluwer Academic Publishers.

47. Sonderegger R C, 1978, 'Diagnostic tests determining the thermal response of a house', in *ASHRAE Transactions*, Vol. 84 (1), pp. 691–702.

Calculation methods and parameter estimation

48. Meier A K and Busch J, 1988, 'Testing the accuracy of measurement-based building energy model with synthetic data', in *Energy and Buildings*, 12, 77–82.

49. Rabl A, 1988, 'Parameter estimation in buildings: methods for dynamic analysis of measured energy use', in *Journal of Solar Energy Engineering*, 110, pp. 52–66.

50. Subbarao K, Burch J and Jeon H, 1986, *Building as a Dynamic Calorimeter: Determination of Heating System Efficiency, Summer Study on Energy-Efficient Buildings 1986, Santa Cruz, California*, Report SERI TP254-2947, Golden, CO, Solar Energy Research Institute.

51. Busch J F, 1986, 'A comparison of building thermal performance models using measured data', in *ASHRAE Transactions*, 92, Paper No. 3011, pp. 537–549.

52. Jansen J E, 1980, 'Application of building thermal resistance measurement', in *ASHRAE Transactions*, 88(2).

53. Wilson N W, Wagner B S and Colbourne W G, 1985, 'Equivalent thermal parameters for an occupied gas-heated house', in *ASHRAE Transactions*, 91(2), pp. 1875–1884.

54. Agami Reddy T, 1996, 'Status of residential short-term performance monitoring methods' in, *Proceedings of the 1996 International Solar Energy Conference*, 359–367. New York: American Society of Mechanical Engineers.

55. MacDonald M and White D, 1990, 'Explanatory investigation of a new modelling approach and an energy index for multifamily buildings in the Pacific Northwest', in *Conference proceedings of ACEEE summer study on energy efficiency in buildings, Pacific Grove, CA(USA) 26 Aug–1 Sep 1990*. Washington: American Council for an American Efficient Economy, 1990.

56. Subbarao K, Burch J and Handcock C E, *How to Accurately Measure the Building Load Coefficient*, Report SERI/J254-0405, Golden, CO, Solar Energy Research Institute.

57. Subbarao K, 1985, *Thermal Parameters for Single and Multizone Buildings and Their Determination from Performance Data*, Report SERI/TR253-2617, Golden, CO, Solar Energy Research Institute.

58. Källblad K and Adamsson B, 1994, *The BKL-Method, A Simplified Method to Predict Energy Consumption in Buildings*, Document D8:1984, Swedish Council for Building Research, Stockholm, Sweden.

59. Kaplan M, Jones B and Jansen J, 1990, 'Model calibration with monitored end-use data', in *1990 ACEEE Summer Study on Energy Efficiency in Buildings*, Vol. 10, Washington DC, American Council for an Energy Efficient Economy.

In situ evaluation of UA and gA values

60. European Committee for Standardisation (CEN), 1997, *Building Components and Elements – In Situ Measurements of the Surface-to-Surface Thermal Transmittance*, prEN 12494, CEN, Brussels.

61. European Committee for Standardization (CEN), 1996, *Thermal Insulation – Determination of Steady-State Thermal Transmission Properties – Calibrated and Guarded Hot Box*, EN ISO 8990 (European Norm), CEN, Brussels.

62. Alexandre J (ed.), 1998, *La Maison PLEIADE – Conclusions*, Jambes (in French).

63. BIN-IBN, 1989, *Thermal Insulation of Buildings – Thermal Characteristics of Buildings*, NBN B62-301, BIN-IBN, Brussels (in Dutch and French).

64. BIN-IBN, 1987, *Calculation of Heat Transmission Coefficients Through Walls*, NBN B62-002, BIN-IBN, Brussels (in Dutch and French).

65. BIN-IBN, 1986, *Calculation of the Heat Losses of Buildings*, NBN B62-003, BIN-IBN, Brussels (in Dutch and French).

66. Belgian Building Research Institute (BBRI), 1997, *Study and Monitoring of the PLEIADE Dwelling – Final Report 1994–1996*, BBRI, Brussels.

67. Somogyi Z (1998), *In Situ Evaluation of the Thermal Characteristics of Building Components and Buildings Including Comparison with Predicted Performances*, PhD thesis, UCL, Louvain-La-Neuve.

68. Akander J, Jóhannesson G and Norlén U, 2001, *The Experimental Protocol of Euro-Class – Level B – 2001-01-10*, Divison of Building Technology, Department of Building Sciences, KTH, Stockholm.

69. Bloomfield D, *et al.*, 1988, *An Investigation into Analytical and Empirical Validation Techniques for Dynamic Thermal Models in Buildings, Final Grant Report*, Science and Engineering Research Council, Swindon (UK).

70. Judkoff R and Neymark J, 1995, *Building Energy Simulation Test (BESTEST) and Diagnostic Method*, IEA Annex 21.

71. Vandaele L and Wouters P, 1994, *The PASSYS Services: Summary Report*, EUR 15113 EN, European Commission, Brussels.

72. Bloem J J (ed.), 1994, *System Identification Applied to Building Performance Data*, EUR 15885 EN, European Commission, Brussels.

73. Martin S, Wouters P and L'Heureux D, 1996, *Evaluation an Application of a Simplified Method for the Certification of Non-Occupied Buildings*, Report prepared for the EC Joint Research Center, Ispra under Contract No. EU 11297-95-10 F1EI ISP B.

74. Stymne H and Eliasson A, 1991, 'A new passive tracer gas method for ventilation measurements', in *Proceedings of 12th AIVC Conference – Air Movement and Ventilation Control within Buildings*, Vol. 3, 1–16, Ottawa.

75. Boman C A and Stymne H, 1996, *PFT – Measurements of Ventilation in European Dwellings*, Report prepared for the EC Joint Research Center, Ispra under Contract No. EN 9505127 T.

76. Helcke G A and Peckham R J, 1991, 'Energy, the computation of degree-days for use as a climatic severity indicator', in *Eurostat Energy Statistical Tables*, EUR 13697 EN, Commission of the European Communities, Luxembourg.

77. European Committee for Standardization (CEN), 1993, *Thermal Performance of buildings – Calculation of Energy Use for Heating – Residential Buildings*, prEN 832 (European Norm), CEN, Brussels.

CHAPTER 2

Experimental methods for the energy characterization of buildings

JAN AKANDER AND GUÐNI JÓHANNESSON

Division of Building Technology, Department of Building Sciences, KTH, Stockholm

DEFINITIONS

Residential building/apartment: A residential building or apartment may be composed of one or more spaces enclosed by walls, ceilings and floors. The building must be situated above the ground and be utilized as a dwelling. At least 70% of the space within the envelope must be used for this purpose. This volume should be heated to at least $X°C$ during Y hours of the year. (Values X and Y are to be determined on a national level).

Energy use: Any form of fuel or electricity, which is actively supplied to the building with the intent of conditioning the living space, heating water and operating energy systems and household appliances. Energy use is given on an annual basis, kWh/(year · building unit).

Global energy use: Total energy use of the whole residential building or apartment. This value can be actual or normalized.

Specific energy use: Energy use with an explicit function, which, when taken with the other specific energy uses, adds up to the total energy consumption of a whole residential building. The types of specific energy use are space heating, space cooling, tap-water heating, appliances and external energy. The value can be actual or normalized.

Energy conversion system: Any system or device in the building with the purpose of converting fuels or energy into work or heat.

Energy conversion system efficiency: The ratio between delivered and supplied energy of an energy conversion system.

Supplied energy: Energy (or an equivalent quantity) that is actively supplied to the conversion systems of the residence. Supplied energy is often referred to as billed energy.

Delivered energy: Energy that is delivered into the residential living space through energy conversion systems.

Actual energy use: Energy use that was observed when the building was utilized in its natural condition for a certain period of time.

Normalized energy use: Actual energy use that is adjusted to a set of reference conditions.

Standard conditions: A set of variables that have pre-scribed values.

Appliances: Devices used for converting one form of energy or fuel into useful energy (heat or work).

Building envelope: All building components that separate the external and internal environments.

Solar aperture: Openings in the building envelope that permit solar irradiation.

Tap warm water: Tap water that has been heated.

Heat-loss factor: Heat flow coefficient from the heated space to the external environment (EN ISO 13790).[1]

Heat gain: Quantity of heat generated within or entering into the heated space from heat sources other than the heating system (EN ISO 13790).[1]

Internal heat gain: Quantity of heat generated within the heated space from heat sources other than the heating system.

Solar gains: Heat gains due to passive solar irradiation.

Heat losses: The amount of heat transferred from heated space to the external environment by transmission and by ventilation, during a given period of time (EN ISO 13790).[1]

External temperature: The temperature of the external air (EN ISO 13790).[1]

Internal temperature: The arithmetic average of the air temperature and the mean radiant temperature at the centre of a room (internal dry resultant) (EN ISO 13790).[1]

Set-point temperature: The design internal temperature (EN ISO 13790).[1]

Thermal zone: Part of the heated space with a given set-point temperature, throughout which the internal temperature is assumed to have negligible spatial variations (EN ISO 13790).[1]

Unheated space: A room or enclosure that is not part of the heated space (EN ISO 13790).[1]

Utilization factor: A factor reducing the total monthly or seasonal gains (internal and passive solar) to obtain part of the useful gains (EN ISO 13790).[1]

Metabolic heat: Heat dissipated from living beings.

THE EXPERIMENTAL PROTOCOLS

This chapter presents the experimental protocols that are to be used to gather information together with the methods of processing the information to prepare values for normalization prior to the rating procedure. It is important to note that the protocols provide the framework for what is to be taken into consideration within the collection and processing of energy values. How this is to be done for individual buildings cannot be specified in the form of various case studies: the case combinations would be endless. Chapter 5 illustrates eight real-scale applications and how these have been dealt with. The framework leaves certain aspects to be determined at a national or regional level. The motivation for this is that information on building technology, end-user behaviour, statistics and informative building codes or methods, and customs (traditions) are better treated within a local framework. The idea is that the framework can be adapted to suit the national conditions in the most appropriate way.

The need for experimental protocols

Rating of buildings by means of energy use is not unusual. There exist a wide range of various rating methods; see for example the review by Kotsaki and Sourys.[2] The rating methods can be customized for residential or commercial building only or they may be applicable to both of these. The majority of methods make use of descriptive models. In other words, a fundamental feature prior to the rating procedure is that technical data on the building is assessed and implemented in a model, which, by means of simulation, gives the thermal performance of the building. The simulation makes use of some reference climate data and standardized indoor conditions for the purpose of evaluating the thermal performance of the building, subject to these conditions. The simulated performance can then be compared with that of other buildings, subject to the same conditions – how good is the thermal performance of the considered building, and why is it behaving in this way? The descriptive models allow energy flows to be traced, and action plans and ratings can be derived from these results.

This type of procedure is common, but how well does the descriptive model simulate the true energy use under prevailing conditions? Research indicates that simulated results based on assumed conditions, for example indoor activities and the living patterns of occupants, can give quite poor agreement with the actual energy use. Now, if the rating of a residence is to be performed, is it reasonable for it to be based on a simulated building with standard indoor conditions and occupants, a reference climate and technical data assessed on the energy conversion systems and a building envelope usually based on plan drawings and audit? The answer is dependent on the purpose of the rating.

The purpose of a rating is specified to classify buildings by their energy use. What affects energy use can be reduced to a set of variables and parameters. The dominant ones are as follows:

- the efficiency of energy conversion systems
- the characteristics of the building envelope
- the external environment
- the behaviour of the occupants, who explicitly control and affect the internal environment.

The experimental protocol is intended to serve as a platform for identifying the effect that these variables and parameters give rise to in terms of energy use, and for determining the magnitude of the specific energy use. The output of the protocol is intended to provide an input to the rating scheme with energy as a basis: the fundamental unit must be a thermal unit, i.e. energy expressed in Joules or equivalent units (kWh for example). The output of the protocol must also be related to the energy that is actually used by the occupants and the building as a whole. This is a fundamental criterion within the classification scheme: it is not a set of pre-determined conditions that will determine the outcome because actual observed energy use must be assessed with an experimental protocol. This energy will then be normalized with reference to the outdoor conditions that have prevailed and to the building size. Normalization is used to make variables usable in various comparison scenarios.

From what has just been discussed, it is clear that the strategy used for designing the protocol must be based on an experimental point of view: the actual energy used over a period of time is the true input to the protocol. The only observations, or measurements, that are carried out over time without an active monitoring campaign are reflected in energy bills. These give the values of actual energy use, and form the heart of a truly experimental protocol.

The aim of the experimental protocol

The aim of the experimental protocol is:

- to determine actual energy use and identify its constituents
- to assess specific energy use, which can be normalized.

The actual energy use is reflected in energy bills or other energy records, which cover a period in the past. The frequency of the bills is an important factor when it comes to interpreting why the energy is used in a certain way and what the energy is used for. As long as one is satisfied with knowing how much energy was actively supplied to the building for a period of one year, bill information is perfectly adequate. However, this is not the case if the intention is to compare the energy use of one building with that of another. Or if one wishes to examine the influence of weather or the location of the building – or what the effect is of other occupants in the same residence. The actual energy use is primarily dependent on the external climate, which nothing can control. For this reason, it is convenient to relate energy use to a climate that is known and applies to a set of buildings: a reference climate. This climate will influence energy use and, by definition, a building subjected to the reference climate will lead to an energy use that is called normalized energy use. However, the prevailing internal conditions of the actual and reference climates are assumed to be the same.

It is when determination of normalized energy use is necessary that information and data from the considered building, its occupants and the related bills have to be assessed and analysed. In order to create a method for this procedure, an idealized model including definitions and assumptions has to be formulated. These are the focus of the following sections.

Because normalization will primarily be based on the conditions of the external environment, it is climate-dependent energy use that will be a key issue in the analysis. Primarily space-heating and space-cooling energies will differ for the prevailing and reference climates. Thus, special attention is given to these two entities, since they particularly influence the difference between actual and normalized energy use.

The idealized model of energy flow in buildings – definitions

A model of energy flow in a building is needed as the basis for analysis. The analysis could be performed by following a quantity of thermal energy 'from the cradle to the grave', as this quantity changes forms over time. The metaphor 'from the cradle to the grave' is intended to describe the supply of energy from the stage of its being primary energy to the stage where the energy has been used by an end-consumer (Figure 2.1).

Primary energy is a term that involves energy resources and energy conversion systems on a national level. The experimental protocol will not explicitly deal with primary energy, although it may provide information that is valuable for analysis of primary energy use. The experimental protocol focuses more on secondary energy, in other words, the use of energy that is supplied from primary energy conversion systems. Within this context, this is called supplied energy.

SUPPLIED ENERGY

Supplied energy is what can often be called billed energy, because energy supplier companies measure the quantity of energy supplied to the end-user (the resident) and bill that quantity. The protocol aims to analyse energy use from the point at which supplied energy is metered to the point where energy exits the residence.

Supplied energy is often metered, after which the energy is often converted to another form by means of a supplied energy conversion system installed in the building. The energy conversion system may be a heating system, for example where the heat content of natural gas is converted to heat. Alternatively, it may be an electric cooker, which converts electricity to heat, or it may be a heat pump that makes use of electricity to generate a cooling thermal load. What is common to all energy conversion systems is that they have performance indexes, or efficiencies, that indicate how well the conversion process takes place.

From the point of view of the protocol, the supplied energy is vital input data, since it provides a basis for estimating the energy that is delivered from the energy conversion systems into the building, the so-called delivered energy. Delivered energy is in fact what is commonly called net energy.

DELIVERED ENERGY AND SPECIFIC ENERGY

The output from supplied-energy conversion systems is thermal energy. This energy is called delivered energy, which is needed for creating the thermal indoor environment, for operating energy conversion systems and for fulfilling the needs and functions that are required by the occupants. To illustrate the difference between supplied energy and delivered energy, the example of a heat pump can be used. Suppose the heat pump has a coefficient of performance (COP) corresponding to 3. If the supplied energy, in this case electricity, is 1 kWh, then the heat delivered from the heat pump is 3 kWh.

Energy delivered from residential energy conversion systems is expressed in thermal units (kWh) and provides the basis for comparison scenarios for buildings that are going to be rated. The task of the experimental protocol is to determine delivered energy from the energy conversion systems in the actual building and to normalize this quantity.

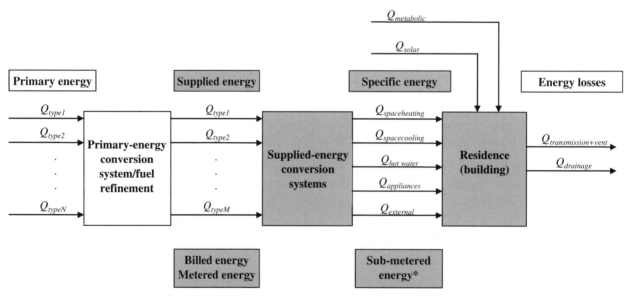

Figure 2.1 An idealized view of energy flow. The methodology is limited to analysis of the energy flow from the point where energy is billed to end-user customers (the areas shaded in grey). Q denotes energy flow and the asterisk (*) the areas that are not often metered

Within the frame of the experimental protocol, delivered energy is divided into five utilization types. These so-called *specific energies* are:

- **Energy for space heating:** Thermal energy delivered from space-heating units with the purpose of heating the residential space (delivered energy).
- **Energy for space cooling:** Thermal energy delivered from space-cooling units with the purpose of cooling the residential space (delivered energy).
- **Energy for tap-water heating:** Thermal energy delivered from tap-water heating units to the water, with the purpose of heating tap water from the temperature it had on entering the building to the temperature it has after exiting the heating units (delivered energy).
- **Energy for internal appliances:** Equivalent thermal energy delivered from domestic appliances and lighting within the residential envelope (delivered energy).
- **Energy for external appliances and spaces:** Equivalent thermal energy delivered from appliances and lighting outside the insulated building envelope. This energy is provided separately, but not included in the heat balance of the building. In contrast, external energy that is delivered from appliances and lighting within the insulated building envelope, but not within the space of the considered apartment, is included in the rating of the building.

There are two specific energies that are not accounted as being delivered energy from residential conversion systems. These are commonly part of the heat gain, which in essence is the heat delivered into the living space, but which is not controlled, or is subject to limited control, in terms of thermal climate. This energy is not purchased, and is represented as:

- **Energy from solar irradiation:** The net solar energy that is transmitted through apertures into the residential space and/or for solar heating of tap water.
- **Energy from metabolism:** Heat dissipated from occupants and other living creatures within the living space of the residence.

Various forms of specific energy consumption can consist of a single fuel source or a combination of different fuel sources. Table 2.1 displays possible combinations of fuels and their use.

ENERGY FOR SPACE HEATING

The energy required for space heating is heat that is delivered from heating units to the heated space during the heating season. For the period of time considered, delivered energy from the heating units is formulated such that

$$Q_{spaceheating}^{delivered} = \sum Q_{heating\ unit}^{delivered} \qquad (2.1)$$

The billed energy for the same period corresponds to heat supplied to the heating system and to the appliances, expressed as follows

$$Q_{spaceheating}^{supplied} = \sum \frac{Q_{spaceheating}^{delivered}}{CF \cdot \eta_{sh}} \qquad (2.2)$$

Here η_{sh} depicts the heating season efficiency of each conversion system (heating system) and CF denotes a conversion factor. The conversion factor relates the thermal unit kWh to the billed energy units. From this point on, supplied energy will be expressed using the unit kWh.

ENERGY FOR SPACE COOLING

The required energy for cooling of spaces is the sum of specific energies that affect the heat balance of the space. Delivered space cooling energy corresponds to

$$Q_{spacecooling}^{delivered} = \sum Q_{cooling\ unit}^{delivered} \qquad (2.3)$$

The billed energy for the same period corresponds to the heat supplied to the heating system and to the appliances, expressed as follows

$$Q_{spacecooling}^{supplied} = \sum \frac{Q_{spacecooling}^{delivered}}{CF \cdot \eta_{sc}} \qquad (2.4)$$

Here η_{sc} depicts the seasonal efficiency of each conversion system and CF denotes a conversion factor. The conversion factor relates the thermal unit kWh to the billed energy units.

ENERGY FOR TAP-WATER HEATING

The required energy for tap-water heating is the sum of the specific energies that affect the energy balance of the tap water. For the period of time considered, the required energy $Q_{hotwater}^{required}$ is formulated such that

$$Q_{hotwater}^{required} = \sum Q_{hotwater} + \sum Q_{solarhotwater} \qquad (2.5)$$

The first term involves active delivery of heat (billed), whereas the second term involves the case where solar collectors are used for tap-water heating.

Table 2.1 Various fuels and forms of specific energy consumption. The symbol (X) implies that the occurrence is very limited

Energy source	Space heating	Space cooling	Water heating	Appliances
Natural gas	X	(X)	X	X
Electricity	X	X	X	X
Fuel oil	X		X	X
Liquefied petroleum gas	X		X	X
Kerosene	X		X	X
Solid fuels	X		X	X
District heating/cooling	X	(X)	X	
Solar radiation	X		X	(X)

The delivered energy is that which is actively delivered to the tap water by means of heating coils, omitting the part that is delivered by solar collectors:

$$Q_{hotwater}^{delivered} = \sum Q_{hotwater} \qquad (2.6)$$

The billed energy for the same period corresponds to the sum of the energy supplied to the tap-water heating units:

$$Q_{hotwater}^{supplied} = \sum \frac{Q_{hotwater}}{CF \cdot \eta_{hw}} \qquad (2.7)$$

Here η_{hw} depicts the seasonal efficiency of each conversion system and CF denotes a conversion factor. The conversion factor relates the thermal unit kWh to the billed energy units. Note that the latter term may derive from a combination of systems, such as a boiler and a heat pump.

ENERGY FOR APPLIANCES AND LIGHTING

The required energy for appliances and lighting is the sum of the energy dissipated from units within the building envelope of the residence or apartment. For the period of time considered, the delivered energy from appliances and lighting corresponds to

$$Q_{appliances}^{delivered} = \sum Q_{appliances\ and\ lights}^{delivered} \qquad (2.8)$$

The part of the bills that denotes supplied energy to appliances is

$$Q_{appliances}^{supplied} = \sum \frac{Q_{appliances}}{CF \cdot \eta_{ap}} \qquad (2.9)$$

Here η_{ap} depicts the seasonal efficiency of each conversion system and CF denotes a conversion factor. The conversion factor relates the thermal unit kWh to the billed energy units. Most household appliances, lighting and other appliances that are used for operating systems in the building make use of electricity and some cookers make use of gas. The seasonal efficiency is therefore very close to unity. However, there are some appliances for which this is not the case. An example of this is when an open fireplace is used for aesthetic reasons, not for the sake of space heating. The efficiency may in this case be close to zero.

ENERGY FOR EXTERNAL APPLIANCES

The required energy for external appliances is the sum of the energy delivered from devices outside the building envelope. These are not considered to affect the heat balance of the building. For the period of time considered, the required energy $Q_{external}^{required}$ is formulated such that

$$Q_{external}^{required} = \sum Q_{external} \qquad (2.10)$$

The delivered energy outside the building envelope is

$$Q_{external}^{delivered} = 0 \qquad (2.11)$$

Hence, the billed energy corresponds to the supplied energy, which here is

$$Q_{external}^{supplied} = \sum \frac{Q_{external}^{required}}{CF \cdot \eta_{ex}} \qquad (2.12)$$

In the case of multi-family buildings, there may be a certain energy use within the envelope but outsidethe apartment

considered. Examples of these are energy for operation of systems, such as central fans and elevators, heating of common spaces such as staircases, common laundries and lighting in corridors. Although the distribution of this energy may be difficult to assess, estimations should be made. The delivered energy for this purpose must be included as a delivered energy to the apartment. The distribution of external energy within the multi-family building envelope is to be determined on a national level. Within this project, the common external energy is divided among the apartments on the basis of the part of heated floor area that the apartment has in relation to the total heated floor area of apartments in the building.

Delivered energy inside the building envelope is

$$Q_{external}^{delivered} = \sum Q_{external\ devices\ and\ spaces} \qquad (2.13)$$

ENERGY FROM METABOLISM

Human beings and warm-blooded pets (occupants) dissipate heat to their nearest environment. When a heat balance has to be calculated, this heat should be taken into account. The dissipated heat can be estimated to correspond to

$$Q_{metabolic} = \sum_{i=1}^{n} \Phi_i \cdot t_i \qquad (2.14)$$

where Φ_i and t_i are the dissipated thermal power (W) and the period of presence (h) of occupant i, and n is the number of occupants. The period of presence is obtained from an audit record.

This energy is not included in the rating entities, but will affect the magnitude of the energies obtained from the assessment procedures.

ENERGY FROM SOLAR IRRADIATION

Assessment of solar radiation that is transmitted into the building requires information on the geometry, orientation, shading and topography of the glazed envelope elements. Solar apertures A_s (the effective areas of the elements) represent these parameters. The transmitted solar energy during a period of observation is calculated as follows:

$$Q_{solarheat} = \sum_j \left(q_{s,j} \cdot \sum_n A_{s,n,j} \right) \qquad (2.15)$$

Here, $q_{s,j}$ (Wh/m^2) denotes the total global solar radiation on the surface $A_{s,n}$ (m^2) in the orientation j. A requirement is that data on global solar radiation on surfaces for different orientations are available for the period of observation. More information is given in Appendix 2.

This energy is not included in the rating entities, but will affect the magnitude of the results of the assessment procedures. Energy from solar irradiation is primarily used for normalizing space-heating and space-cooling energies, and is indeed a part of the heating requirement of the building.

Another energy input that solar radiation gives rise to is heating of tap water. This entity is considered to be a part of the energy requirement for tap-water heating. However, this energy will not be included in the rating procedure and will in most cases be considered a part of the specific energy called Q_{solar}, which is omitted as an input to the rating procedure. Therefore,

$$Q_{solar} = Q_{solarheat} + Q_{solarhotwater} \qquad (2.16)$$

Although Q_{solar} is a *delivered* energy into the residence, it will not have the *delivered* index included.

Climate-dependent and climate-independent variables

When energy use is analysed, it is convenient to make assumptions on whether or not the *specific energy use* is climate-dependent or climate-independent. A climate-independent variable will not change in time (or space) as a result of changes in the internal and external environments. The specific energies that are considered to be *climate-independent* are those that are directly affected only by occupancy behaviour and needs:

- **Energy supplied to appliances and lighting.** Although use of certain appliances may be more or less seasonally dependent, such effects will be neglected. For example, lighting is in general dependent on solar radiation and time of year.
- **Energy for tap-water heating.** This energy is assumed to be evenly distributed throughout the year or season. This may not be completely true because, for example, showering may be more frequent during the summer season. The temperature of the incoming tap water may also vary seasonally.
- **Metabolic heat dissipated by people and household pets.** This is assumed to be the same throughout the year, provided that the occupancy rate within the residence is the same. Metabolic heat is somewhat dependent on the temperature of the indoor climate, but within this context will be assumed to be constant.
- **Energy supplied to external devices.** This is assumed to be climate-independent on a seasonal basis. Although this is not entirely true, because, for example, of the frequent use of car heaters during colder periods during the winter season, this entity is assumed to have a poor correlation with climate.

Climate-dependent variables are those that are affected by climate: specific energy use increases when the climate becomes more severe:

- **Space heating.** Energy delivered from the heating units of the residence increases as the climate gets colder. The increase and decrease with climate variations can be regarded as being linearly dependent.
- **Space cooling.** Energy for space cooling is, unlike space heating, non-linear. However, the energy used for cooling does increase when the outdoor climate becomes warmer and with increased solar radiation.
- **Solar energy into the residential space.** This is climate-dependent, but can to some extent be controlled by means of shading devices.

- **Energy supplied to external spaces, for heating and cooling purposes.** This is considered to be climate dependent.

PROPOSAL OF TWO PROTOCOLS

The access to observations on energy use of a building in time depends on the periodicity and the quality of the data included on bills. Also, there may be energy use that is not billed (non-billed energy use). This may not only be dependent on traditions and lifestyle in national or regional terms, but it may also differ between energy suppliers and be covered by different agreements between companies and the energy suppliers.

Apart from the quality of the supplied energy, the data on other energy parameters is often limited. In addition to fuel use, the data need to perform an energy analysis of a building are primarily:

- external climate
- internal climate
- occupant behaviour
- efficiency and utilization patterns of energy conversion systems.

In order to perform a more-or-less 'exact' analysis of energy flow during the current year and a reference year, a whole year of high-frequency extensive monitoring would have to be performed. This is not practical. A decision has to be made, therefore, in which the inaccuracy of the results is weighed against the cost of labour and resources. In addition, there is the question of the willingness of the customer to purchase, not only for a rating score, but also information on the reasons for the score. This additional service requires more resources and will increase costs. For this reason, two separate protocols have been designed fulfilling the same requirements but with different degrees of detail. The costs, the resources, the delivered information (service) and the level of accuracy of the results will allow the customer (occupants) to choose a protocol. The results generated by each protocol are compatible in rating since they have a similar experimental platform.

The two protocols are called:

- the Billed Energy Protocol (BEP)
- the Monitored Energy Protocol (MEP).

OVERVIEW OF BEP AND MEP

Assessment of the information that is needed to determine the energy use of a residence may require immense resources and time, thus leading to expensive long-term monitoring campaigns. For this reason, the two protocols BEP and MEP are available. What distinguishes these two is the level of detail in data and information, both in the input and in the output. The person who desires to rate a building makes a choice, which will depend on what input data and information is available, whether procedure criteria are fulfilled, what types of results are desired and how much

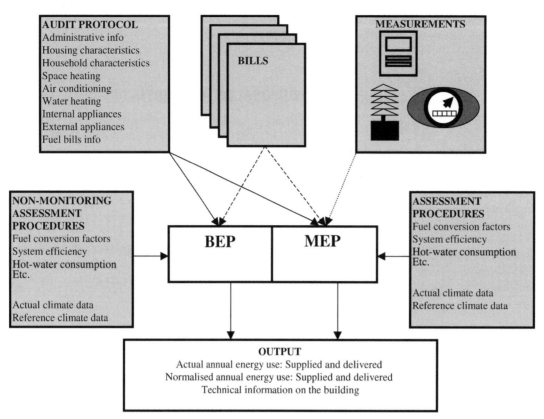

Figure 2.2 The two protocols BEP and MEP, which can be used for assessing the energy supplied and delivered and the technical information. Each protocol requires a different amount of input data amount and different assessment methods. The input sources are shaded

they are willing to pay for this service. The two protocols are shown in Figure 2.2.

The simpler protocol, BEP, is conducted within four man-hours per residence, provided that the input data fulfil several criteria. The output consists of the actual and the normalized annual supplied and delivered specific energy use. Normalization of energy use is only performed with regard to the outdoor climate, because indoor temperatures are unknown. To decrease the cost of the rating process, the only measurements that will be made will be reading meters that already are installed in the building. The fact that measurements are not conducted will lead to an increased uncertainty in the results.

MEP makes use of bill records, but is more flexible since the protocol involves measurements of climate variables and sub-metering within the building to enhance the accuracy of the input data. MEP is also recommended in the case where the criteria for BEP prevent application of the simpler protocol. MEP output will in most cases be more reliable than that from BEP. Moreover, normalization of energy use is performed with the reference outdoor climate and the indoor climate taken into consideration. The monitoring period is at least two weeks long, although from an energy point of view two weeks is considered as a short-term measurement. MEP may involve a concentrated monitoring scheme, lasting for over 10 weeks. The duration of the monitoring depends on which service, aside from rating, is purchased. Technical information from MEP is important to document, because this information can be used to establish default values for BEP. An example of this is

the boiler efficiency as a function of fabrication, type and age.

The protocols are summarized in Tables 2.2 and 2.3, which give an overview of the general issues for each procedure. For more detail, please refer to the following sections in this chapter, which give more detail on each procedure.

BILLED ENERGY PROTOCOL (BEP)

The Billed Energy Protocol (BEP) is the simpler of the two experimental protocols. Relying heavily on bill information and earlier records, this protocol receives its name based on the source of the most important information that is assessed. Also using information obtained from an audit at the residence, this method does not explicitly make use of any monitoring campaigns. The protocol is to be performed at a minimum cost. However, in order for the outcome of the protocol – the results of energy use – to be adequate in terms of accuracy, there are several requirements that must be fulfilled in order for BEP to be executed. If these requirements are not met, the other protocol (MEP) should be applied.

Strategy

BEP relies on information from bills and the audit as input. The procedures within BEP that are used to derive specific energy will, in cases where there may be lack of information, make use of estimations or default values. Because it

Table 2.2 An outline of BEP and MEP

Protocol	BEP	MEP
Actions	Audit at site	Audit at site
	Use of energy bills	Use of energy bills
		Use of monitored data
Measurements	Readings of existing gauges during the audit occasion	Metering of supplied or delivered energy
		Measurement of energy system efficiency
		Indoor and outdoor temperature measurements
		Optional solar radiation measurements
		Optional heat loss factor measurement
		Optional measurement of ventilation rates, air-tightness
Duration of audit	1 – max 4 hours	2 – max 8 hours
Duration of measurements	None	A minimum of 2 weeks – about 10 weeks, depending on building type and purpose

Table 2.3 Input and output variables for the two protocols

Protocol	BEP	MEP
1. Supplied energy	Billed quantities	Billed quantities
		Measured (min 2 weeks)
2. Efficiencies	Default values or from previous tuning records	Measured
3. Specific energies	Estimated annual use	Measured (min 2 weeks)
		Estimated annual use
4. Delivered energy	Estimated on basis of 1, 2 and 3 above	Estimated on basis of 1, 2 and 3 above
5. Outdoor temperature	Closest climate station	Measured (min 2 weeks)
		Climate station for annual data
6. Indoor temperature	No	Measured (min 2 weeks)
		Mean value assumed same for season of measurement
7. Solar radiation	Closest climate station	Measured (min 2 weeks)*
		Climate station for annual data
8. Normalization of outdoor climate	Necessary	Necessary
9. Normalization of outdoor and indoor climate	No	Necessary
10. Determination of residential heat loss factor	No	Optional[†]
11. More detailed technical building data	No	Optional[‡]

* Optional, depending on whether or not solar radiation has a large influence on the heat balance of the building during the monitoring period.

[†] If possible, depending on the data quality and the season during which measurements were performed, and whether the customer desires this service.

[‡] Extra services such as measurement of rates of ventilation, air-tightness of the envelope, indoor air quality, thermal bridges, U values for specific envelope components and lighting characteristics.

depends on the quality of the bills, the billing periodicity, the availability of conversion system tuning records, sub-metered values and reliability based on information supplied by the end-user, the outcome of the procedures should always be considered to be an estimation. BEP does not strive to evaluate the magnitude of the inaccuracy that the estimations involve: this inaccuracy will not only vary depending on which building is considered; it will also depend on the systems, end-user behaviour, bill quality, availability of climate data and the judgements of the auditor. It will also depend on default values or national standard calculation or statistical methods, which are used as estimation procedures when data for the considered building is not assessable.

A step-by-step illustration is given below, summing up the major steps involved in BEP. The steps are described in more detail in succeeding sections. These involve the following:

- **Pre-arrangements:** The auditor informs the occupants about the experimental protocol, how it works, limitations etc. A quick check is made on whether or not the residence fulfils the BEP requirement list.
- **Collection of billed data:** Series of bills are collected and the quality is evaluated.
- **Collection of non-billed data:** Series of non-billed quantities are collected (on the basis of information supplied by the end-user or a third party).
- **The audit:** General information is assessed for the building. For this purpose, a checklist is provided for the auditor. The general information covers aspects such as:
  - administrative and technical data on the building itself
  - technical data on the energy conversion systems and energy distribution within the residence
  - end-user behaviour.

- **Analysis of bills:** The analysis will take into consideration:
  - energy supplied to each energy conversion system
  - an estimate of the delivered energy from each energy conversion system
  - an allocation of each delivered energy to the correct specific energy, special attention being focused on the specific energies for space heating and space cooling.

    Making assumptions on, for example, seasonal energy use can ease the analysis of bills.
- **Optional services:** If the quality of the input data is sufficient, other informative services may be provided, but are not required, within BEP. Such a service can be to estimate the heat loss factor of the building.
- **Results:** The outcome of BEP is the following:
  - an annual record of the supplied energy
  - an annual record of the estimated values for the five specific energy types.

    It is necessary to specify how the values have been obtained and that the calculations are described in the accompanying list of footnotes.

Pros and cons

BEP is completely based on information that can be acquired from previous bills and from an audit that does not include monitoring. BEP is experimental in the sense that billed quantities are often measured and descriptive calculation models are not used. In fact, the major input data is energy supplied during the period considered.

If the protocol only dealt with supplied energy, the summing of bills would give adequate information on energy use. However, more information and procedures have to be applied if the output of the protocol is to be delivered energy with allocation into the various specific energies. Furthermore, if these values are to be normalized, more input is required.

For the protocol to be affordable, both in time and resources, it has to be simple, but it must also be weighed against result inaccuracy. This means that the procedures within BEP may rely quite heavily on estimations – certainly for allocation of each of the delivered energies to a specific energy category. Moreover, the efficiency of energy conversion systems may not have been measured – alternative methods will have to be used in order to proceed with the assessment process.

In Table 2.4, the limitations of BEP are listed, including some comments as to the reasons for the limitations.

Conditions for the application of BEP

As a result of the design of BEP, it is necessary that certain conditions be fulfilled. These conditions are necessary to limit the uncertainty in part of the input data, which is introduced into the procedures. Should these requirements not be fulfilled, it will not be possible to obtain reliable results. Rating will not be possible according to BEP. The alternative is to proceed with MEP.

Table 2.4 The limitations of BEP

Aspect	Assessment	Comment
Indoor climate	Not considered	The occupant supplies information on indoor temperatures, but this is considered to be somewhat unreliable. Monitoring is not performed because of time limitations. Set-point temperatures, if possible, are assessed during the audit (monitoring is done in MEP)
	Temporary value from the audit or information supplied by the occupant	
External climate	Estimated	Data from the nearest climate station can be used
	Not measured	Discrepancies in micro- and macroclimate are not considered to be greater than inaccuracies in other steps (local monitoring is done in MEP)
Energy conversion system efficiencies	Estimated	Time-consuming and costly
	Not measured	Default values, older tuning records or national procedures can be used (measurements made in MEP)
Sub-metering of specific energy	Estimated	Time-consuming and costly
	Not measured	The largest amount of estimations are made for multipurpose fuel use. The allocation of delivered energy to specific energy requires estimations. Estimations can be made using default values, statistics or prescribed codes. These have to be determined on a national level, to suit the framework for country-specific constraints. The use of estimation procedures introduces errors into the allocation, which will affect later calculations on normalization (sub-metering is done in MEP)
Solar energy and metabolic heat	Estimated	Solar energy and metabolic heat can be calculated from relevant assessed values (solar apertures and occupant presence), but with large uncertainty. Solar energy will require computational procedures (the same applies in MEP)
Non-billed supplied energy	End-user estimation	Non-billed supplied energy information has to be supplied by the end-user or a third party. The reliability of this information cannot be verified. If the annual value exceeds the global annual energy by 30%, MEP must be used (short-period measurement is used in MEP)

THE REQUIREMENTS CONCERNING THE RESIDENCE AND THE OCCUPANTS

- The residence must fulfil the definition of being a residence.
- The building should not have undergone retrofitting or extension or had any changes made to the energy systems during the period of time that is covered by the set of bills being used (primarily the last year). The same applies to the number of occupants, their behaviour and their utilization patterns and the rate of occupancy.

THE REQUIREMENTS CONCERNING THE BILLED AND RECORDED ENERGY CONSUMPTION

- There must be complete sets of bills or readings from gauges that correspond to the annual total energy consumption. The bills have at least to include information on the fuel type, the quantity of fuel and the time of delivery, as well as whether or not the billed values are estimations/predictions on the part of the energy supplier.
- At least two bills for each fuel have to be assessed during the year. If this is not possible, readings of gauges should be made between the seasons. Values read during the audit are considered to be billed values.
- If the bills are based on estimations by the energy suppliers, it is necessary that the billed estimated values and the readings of values do not radically differ at the time of audit. The maximum deviation allowed is 30% on an annual basis.
- For fuels used, but not billed, it is necessary to have recorded data on an annual basis, as for ordinary bills. If this information cannot be verified, the procedure cannot be executed unless this non-billed fuel comprises less than 30% of the total energy consumption.
- The collection of reliable data may include a period of more than a year if relevant input values are to be obtained. For example, fuel oil may only be delivered once a year.
- Conversion factors from fuel types to the unit Wh must be available. The energy content of each fuel type is determined at a national level.

THE REQUIREMENTS CONCERNING AUDIT AND CLIMATE INFORMATION

- The energy auditor must have inspected the site, accessed all premises, read all relevant gauges and completed an accurate audit form. It is assumed that the residence representative provides 'true' behaviour and utilization patterns.
- Reliable climate data must exist for the area in which the building is situated and must be obtainable from the closest climate station. Reference climate data must exist for all regions for calculation of the Climate Severity Index or other climate data that influences the heat balance of the building.
- It must be possible to access databanks or national codes for energy conversion system efficiencies. These lists or codes should contain information on devices, types, rated power/output, age, performance indexes and efficiencies, performance degradation over time, etc. Values taken from these lists or codes are to be used as default values.
- A regional or national statistical correlation must exist between tap-water consumption and hot-water consumption, or information must be available on hot-water use as a function of the number of occupants or the building space. These values can be used as defaults for assessing energy for tap-water heating. National methods or codes can be applied.

Step-by-step description of BEP

PRE-ARRANGEMENTS

When the occupants of a residence have expressed interest in the classification process, the auditor gives them information about the 'Classification', how it works, its limitations, and what is required etc. A quick check is made on whether or not the residence fulfils the BEP requirement list.

COLLECTION OF DATA ON THE SUPPLIED FUELS

The auditor collects a series of bills. The bills can be obtained from occupants or/and energy suppliers. Preferably, the bills should cover the most recent year/years because building performance or user behaviour may change over time.

Series of non-billed quantities are also collected and the end-user or a third party provides information about these. It is preferable that this information can be verified. In the worst case, the end-user will have to make an estimation of the quantity used. The estimation can either be made on a seasonal basis or by providing information on the frequency of use and how much fuel is consumed on each occasion.

The collected data is entered into Form table 1:

- **Row 1:** The type of fuel is noted in the first row. The number of columns depends on the number of fuels used.
- **Row 2:** For billed fuels, whether the billed quantity is estimated or measured is entered.
- **Row 3:** The conversion factor for heat content per unit fuel must be listed, including the conversion units.
- **Rows 4–:** The period and the billed quantity are entered with the thermal unit kWh.
- **Row n:** Non-billed fuel quantities are entered. Since these are not billed, they are considered to be estimates for the corresponding periods.

THE AUDIT

The audit provides an opportunity for the auditor to physically assess general information on the building. For this purpose, an audit form is provided for the auditor. An audit form is proposed and described in Appendix 1, but it is recommended that national audit forms be created to

Form table 1 Supplied energy

SUPPLIED ENERGY	Type 1	Type 2	Type 3
Estimated/Measured			
Conversion factor			
Billed: period 1			
Billed: period 2			
Billed: period 3			
. . .			
Non-billed: period			
SUM (kWh/year)			

make the inquiry more targeted. The general information covers aspects such as:

- technical data on the building
- technical data on the energy conversion systems and the energy distribution within the residence
- end-user behaviour.

The information from the audit is complementary to the bill information and can be used for the purpose of analysing the energy use. An important issue is assessment of the types of energy conversion system that are present in the building, and assessment of the efficiency of these systems. These are entered in Form table 2.

- **Column 1:** A description of the energy conversion system is entered here. Some examples are:
  - boiler for tap water and space heating (mark, age, rate power, etc.)
  - lighting, household appliances and services
  - stoves (type, mark, age, rate power, etc.)
  - portable electrical radiator (type, mark, age, rate power, etc.)
- **Column 2:** Fuel type used by each energy conversion system. Some systems may use two fuel types and the estimated proportions should be noted.
- **Columns 3 and 4:** Winter and summer efficiencies for each system are entered here.
- **Column 5:** Are the efficiencies estimated E, measured M or calculated C according to some norm? One of the alternatives is entered with a comment on which norm was applied. Estimated values should indicate whether or not these are national recommended values or manufacturer values.

ANALYSIS OF BILLS

Analysis of bills can be eased by making assumptions about, for example, seasonal energy use. The analysis will take

into consideration:

- energy supplied to each energy conversion system (Form table 1)
- estimated delivered energy from each energy conversion system (Form table 2)
- Allocation of each delivered energy to the suitable specific energy, special attention being given to the specific energies for space heating and space cooling (Form table 3).

Form table 3 is composed of non-shaded blocks representing delivered energy from each energy conversion system considered. The sum of elements in each row (Period . . .) must correspond to the fuel supplied to the system multiplied by the system efficiency. Each element is an allocation with respect to the specific energy to which the delivery from the system was made. The lowest row for each non-shaded block (SUM) shows on an annual basis the system's contribution to the specific energy.

Of great importance are the shaded rows (Method). The footnote is a number that points to a footnote, which gives more detail and accounts for:

- what assumptions were made (seasonal calculations, climate dependency, etc.)
- which methods were used (national codes, statistical relationships, etc.)
- how the values were obtained/derived (motivations, calculations, equations, etc.).

The footnotes are lists that are separately attached to the table. An important feature is that the footnote information shows calculations and references to methods. The results and calculation procedures are thus transparent.

The last row (SUM (all)) is the sum of all the system sums in each column. These values represent annual specific energies, which will form the inputs to the normalization and rating procedures.

OPTIONAL SERVICES

If the quality of the input data is sufficient, other informative services may be provided, but these are not required within BEP. Such a service can be to estimate the heat loss factor of the building (see sections on MEP). In certain countries such options may be required. For example, in Germany and Switzerland, where energy-use monitoring for individual apartments is required by legislation, the data quality will have to be high enough for this option to be mandatory.

Form table 2 List of energy conversion systems, their efficiencies and how these were assessed

ENERGY CONVERSION SYSTEM	Fuel type	Winter efficiency	Summer efficiency	Estimated/Measured/Calculated
System 1	Type 1			
System 2	Type 2			
System 3	Type 3			
. . .				

Form table 3 Allocation of delivered energy, on an energy conversion system level, to each of the specific energies

Method	Footnote 1	Footnote 2	Footnote 3	Footnote 4	Footnote 5
SYSTEM 1	Space heating	Space cooling	Appliances and lighting	Tap hot water	External appliances/spaces
Period 1					
Period 2					
Period 3					
. . .					
SUM					

Method	Footnote 6	Footnote 7	Footnote 8	Footnote 9	Footnote 10
SYSTEM 2	Space heating	Space cooling	Appliances and lighting	Tap hot water	External appliances/spaces
Period 1					
Period 2					
Period 3					
. . .					
SUM					

Method	Footnote 11	Footnote 12	Footnote 13	Footnote 14	Footnote 15
SYSTEM 3	Space heating	Space cooling	Appliances and lighting	Tap hot water	External appliances/spaces
Period 1					
Period 2					
Period 3					
. . .					
SUM					
SUM (all) kWh/year					

RESULTS

The outcome of BEP is the following:

- an annual record of supplied energy (see Form table 1)
- an annual record of estimated values for the five specific energy types (see Form table 3). It is necessary that how the values have been derived is clear in the presentation.

Processing of data

The analysis of bills is the heart of BEP. Though this is a critical phase, it is difficult to provide guidelines to cover every situation that can arise within the whole building stock. The following sections are therefore more descriptive in manner and highlight some of the issues and problems that may be encountered.

FREQUENCY OF BILLS

Bills may not be periodically distributed in time. When examining a year, it is important that the bills are compiled so that their values coincide within the same year and each season. For this reason, interpolation between the time of bills is needed to find how much energy was consumed during the period of interest.

There are two types of energy use. The first is the climate-independent energy use, which can be considered to be constant on a daily, weekly or monthly basis. It may be different for different seasons. Climate-independent energy use is interpolated or extrapolated on the basis of time. The second type is climate-dependent energy use. Interpolation between various billing periods is conveniently carried out with weighting factors. A convenient weighting factor is the use of heating or cooling degree-days over the period. Energy use is assumed to be proportional to the degree-days of the period considered.

TAP HOT WATER

It is seldom that energy for heating tap water is measured. The reasons are to a great extent that sub-metering is never carried out:

- The heat for tap water may be integrated with the boiler for space heating.
- The quantity of water supplied to the residence is measured, but not its distribution into the hot and cold water circuits.
- Solar collectors, if present, deliver heat on the basis of the outdoor climate.

There are basically two ways of estimating heat delivered for water heating purposes.

- determining the energy supplied to the tap-water heating system
- determining the energy delivered to the tap water.

Supplied energy known

The first of the two ways is to determine how much fuel has been supplied to the energy conversion system. If this is the case, then the heat delivered for tap-water heating corresponds to

$$Q_{hotwater}^{supplied} = \sum \frac{Q_{hotwater}^{delivered}}{CF \cdot \eta_{hw}} \qquad (2.17)$$

When the supplied fuel quantity is known, multiplication by the fuel conversion factor and the seasonal efficiency of the energy conversion system gives the energy delivered for tap-water heating. For BEP data processing, the conversion factor and the seasonal efficiency make use of default values or data delivered from energy suppliers or product manufacturers or data from previous tuning records.

In the case where a fuel is for multipurpose use, sub-metering records will be required. They are seldom available in practice.

Delivered energy known

The second way of assessing the heat delivered to tap hot water requires that the amount of water that has passed through the system has been recorded and, at the same time, that the rise in water temperature has been registered. The mathematical expression is

$$Q_{hotwater}^{delivered} = M_{hotwater} \cdot c_p \cdot \Delta\theta \qquad (2.18)$$

where $M_{hotwater}$ is the mass of water that has passed through the heating devices during time t, the specific heat capacity of water is c_p (4,180 J/kg·K) and $\Delta\theta$ corresponds to the mean increase in temperature.

In the case where the mass flow of water has been recorded, but no temperature rise is available, the temperature rise may be estimated on the basis of the set-point temperature of the water-heating device or from the accumulation tank and the temperature of the supplied tap water. The supply temperature of the tap water can be assumed to be the mean ground temperature of the season.

Sub-metering of this type is not common, but may be present in multi-family buildings with individual monitoring in apartments.

Supplied and delivered energies unknown

When sub-metered values of fuel and hot-water consumption are not available, estimation of delivered energy must be applied on the basis of statistical correlations. These may be derived on a national or regional basis. Examples of these are:

- The ratio between the consumption of domestic hot water and total water consumption should be estimated by statistical means. With the use of the system set-point temperature and the supply temperature of the tap water, the energy delivered to water heating is estimated with Equation 2.18.
- Hot water consumption can be estimated, based on one variable or a set of variables, such as the number of occupants or the size of the household. It is important that the analysis compares this value with the actual tap-water consumption to ensure that the value obtained for the energy is reasonable.
- Tap-water heating energy for a standard family or a single occupant can be used. It is important that the analysis compares this value to the actual tap-water consumption to ensure that the value obtained for the energy is reasonable.
- The entire water consumption is multiplied by the estimated difference in temperature between the waste water and the water supplied.

It is important within these methods to gather information on what influences the patterns of total water use. Total water use can vary to a great extent if watering gardens is frequent, if swimming pools or ponds are used or if the water meter, as well as residential consumption, also measures farming or industrial water consumption.

APPLIANCES AND LIGHTING

The term 'appliances' refers to devices that are used for converting one form of energy or fuel into useful energy (heat or work). For this reason, lighting is, with this definition, also a part of this specific energy.

Appliances will therefore include refrigerators, freezers, clothes washers and dryers, kitchen and household machinery such as cookers, ovens and dishwashers, lights, fans and pumps, and stoves and fireplaces that are not explicitly used for continuous heating or cooling purposes. The energy is delivered within the boundary of the residence (apartment) and will directly affect the heat balance of the residential space.

A common characteristic is that there are no appliance thermostats that control heat (or cooling) dissipation. These devices will continue to dissipate heat even though the space temperature exceeds the set-point temperature of the heating system, often resulting in space temperatures being higher than the set-point temperature of the heating system.

Most appliances used for household functions and operation of the residence use electricity, apart from cookers, ovens, clothes dryers and lighting, which may use gas. For the case where only electricity is used by appliances, bill analysis will give the delivered energy.

STOVES, FIREPLACES AND OTHER HEATING UNITS WITHOUT THERMOSTAT CONTROLS

There are also appliances that do not explicitly use electricity. Examples are stoves and fireplaces that are used more for aesthetic or 'comfort' reasons rather than for continuous space heating. Stoves may use electricity, wood, pellets, coal, kerosene, gas or other fuels. If the quantity of fuel is not billed or recorded, the end-user will be required to make estimations of fuel consumption. This is done in one of the two following ways:

- The end-user is required to give information on the fuel type used, the frequency of use of the appliance per week. The quantity of fuel used per unit time is either estimated by the end-user or is given a national default value. The default value may be based on the type and age of the appliance in question.
- The end-user makes estimations on the seasonal consumption of each type of fuel.

Energy delivered to the space is estimated by means of

$$Q_{appliances}^{supplied} = \sum \frac{Q_{appliances}^{delivered}}{CF \cdot \eta_{ap}} \qquad (2.19)$$

Default values are used for conversion factors and the efficiency of the appliance. If the efficiency of the appliance is available, either from the product manufacturer or from previous tuning records, then these values should be used.

If energy that is delivered from an appliance exceeds 30% of the total delivered space heating, then MEP should be applied. The uncertainty in end-user estimations is the underlying reason for this.

EXTERNAL ENERGY USE

External energy use is different from the other specific energies for three reasons:

- External energy use is divided into two categories. These are (a) energy use in the external environment and (b) energy use within the envelope of the building, but outside the considered space or apartment.
- External energy use does not explicitly affect the heat balance of the considered residence.
- External energy use can be either climate-dependent or climate-independent.

External energy use in the external environment

Energy is often used in the external environment, especially in the case where the property extends farther than the building envelope. Energy used for this purpose may be climate-dependent or climate-independent. What is meant by the external environment is that the energy use is outside the residence envelope. Examples of these are the following:

- a detached heated garage and/or car heaters (climate/season-dependent)
- external lighting (seasonal dependency with the latitude of the building site and the season)
- heaters for sun-courts, greenhouses, porches, roof balconies etc. or other detached spaces (climate-dependent)
- pool heaters and pumps
- grills and gardening appliances (season-dependent)
- ground and gutter heating coils (season-dependent).

External energy use in the external environment is considered to be supplied energy because the energy is being consumed within the property. At the same time, the energy is not being delivered into the residence and is not affecting the heat balance of the residence. It is not dependent on the quality and characteristics of the considered residence, its envelope and systems. The consequence is that this energy should not be a part of the rating scheme.

The general application is that $Q_{external}^{supplied}$ is measured or estimated. Delivered energy will for systems placed in the external environment be $Q_{external}^{delivered} = 0$.

It is the task of the auditor to estimate the parts of that supplied energy/fuels that are used in the external environment if sub-metering is not present for this specific

entity. Depending on how external energy is used, two methods are available for this estimation.

Estimation of energy use for a detached non-residential heated space

Estimation of energy use for a detached non-residential heated space can be done either by estimating how much energy is supplied into the space or by estimation of the heat losses.

Estimating the heat-loss factor of the space involves estimation of the heat losses. The auditor assesses the areas and estimates the U values of the envelope component. The rate of ventilation is guessed. The internal temperature can be measured during the audit and this value can be assumed to be constant during the season. Otherwise, the set-point temperature is read, or the residence representative can provide information on the mean temperature of that space. The heating degree-day concept is used such that

$$Q_{external}^{supplied} = \frac{\left(\sum_j U_j \cdot A_j + \frac{n \cdot V \cdot \rho \cdot c}{3600}\right) HDD}{\eta_{ex} \cdot CF} \qquad (2.20)$$

where
$U_j \cdot A_j$ is the $U \cdot A$ value for the envelope component j (W/K),
n is the rate of air changes in volume V (h^{-1}),
$\rho \cdot c$ is the product of the density (kg/m^3) and the specific heat capacity (J/kg · K) of air,
HDD is the number of degree-days for the considered period (K · days).

Estimation of energy use of external devices/appliances

In the case of devices or appliances in the external environment, and if the energy use is not sub-metered, this energy use has to be estimated. This estimation is preferably made for each season. The auditor is then required to:

- assess the rated power for the device, either by reading labels or manufacturer information, or measuring at site – default values can be used for specific products
- be informed by the residential representative on the frequency of use or the duration of each use
- make a calculation based on the collected information.

The energy use corresponds to

$$Q_{external}^{supplied} = \Phi_{rated} \cdot t \qquad (2.21)$$

where Φ_{rated} (W) is the rated power of the device and t is the utilization time during the considered season. Caution should be taken as to whether or not the rated power refers to supplied or delivered power.

By summing the total duration for the whole year and multiplying this value by the rated power, an estimated energy use is obtained for each device. It is the responsibility of the auditor to judge whether or not the energy of a specific device/appliance can be considered to be negligible. For example, the use of a low-energy light bulb (11 W) for external lighting of a porch will not significantly affect the total energy use of the residence.

External energy use within the envelope

External energy use within the internal environment may sound cryptic. The meaning of this is that energy is delivered within the building envelope, but outside the space of the residence. This energy will to some extent influence the heat balance of the residence.

Examples are mainly applicable to multi-family buildings and houses that are not detached. These are as follows:

- space heating and cooling of common premises such as corridors, staircases, hallways, entrances, storage rooms and garages (climate-dependent)
- energy for lighting of common premises (certain climate dependency)
- elevators (climate-independent)
- pumps, fans and other operational machinery functions that are not explicitly included as internal appliances (climate-independent)
- energy use for operation of common kitchens, bathrooms, laundries, clothes-dryer rooms and saunas (climate-independent)
- hot-water use in common premises (climate-independent).

Assessment of these entities is extremely complicated unless sub-metering is available, which is normally the case. Usually, the costs for these functions are included in residential bills, and the information on energy use ought to be available through maintenance managers or operators.

The general rule is that energy supplied to external devices and non-residential spaces is considered as $Q_{external}^{supplied}$. The energy delivered to these objects is

$$Q_{external}^{delivered} = \sum \eta_{ex} \cdot CF \cdot Q_{external}^{supplied}$$

This quantity is included in the energy that serves as the basis for the rating procedure. However, although it may be climate-dependent to some extent, normalization with regard to climate will not be performed. Neither will this energy be taken into account in the normalization of space heating and space cooling of the considered apartment/residence.

- **Single-family houses:** For single-family houses, external energy use is for most cases included in the residential energy bills. The task of the auditor is to determine what part of the energy use goes directly to the external environment or detached non-residential spaces. This quantity of energy is to be included in the supplied energy, but will be zero in terms of delivered energy.
- **Multi-family buildings:** Buildings that have common premises use energy that is not explicitly applicable to one user or residence (apartment). A common way of financing 'common/shared energy' is by summing all 'common energy use' and dividing this among the occupants according to one parameter or a set of parameters. A model that is widely used for distributing the common energy costs is to divide the floor area of each apartment by the total floor area of all the apartments. The considered apartment is thus billed proportionally to its floor area.

Delivered and supplied energy for seasons

The convenience of partitioning annual energy use, depending on the frequency of bills, into seasonal categories is apparent when the climate-dependent variables for space heating and cooling ($Q_{spaceheating}^{delivered}$ and $Q_{spacecooling}^{delivered}$) are to be estimated and normalized in terms of the external climate. The sum of energy use is on an annual basis formulated such that

$$Q_{annual}^{delivered} = Q_{spaceheating}^{delivered} + Q_{spacecooling}^{delivered} + Q_{hotwater}^{delivered}$$
$$+ Q_{appliances}^{delivered} + Q_{external}^{delivered} \quad (2.22)$$

There may be, in total, four seasons during a year. These are the heating season, the cooling season and two intermediate seasons when space heating and cooling is not applied.

$$Q_{annual}^{delivered} = Q_{heatingseason}^{delivered} + Q_{coolingseason}^{delivered} + Q_{intermediateseasons}^{delivered}$$
$$(2.23)$$

The components of this equation are

$$Q_{heatingseason}^{delivered} = Q_{spaceheating}^{delivered} + Q_{hotwater}^{delivered} + Q_{appliances}^{delivered} + Q_{external}^{delivered}$$
$$(2.24)$$

$$Q_{coolingseason}^{delivered} = Q_{spacecooling}^{delivered} + Q_{hotwater}^{delivered} + Q_{appliances}^{delivered} + Q_{external}^{delivered}$$
$$(2.25)$$

$$Q_{intermediateseasons}^{delivered} = Q_{hotwater}^{delivered} + Q_{appliances}^{delivered} + Q_{external}^{delivered} \quad (2.26)$$

The delivered entities are based on estimations of supplied fuels (billed data and data from the audit) and the efficiencies of the conversion systems. The seasons are either predefined on a national basis (for example based on the regional climate) or on data obtained during the audit. In the case where bills overlap the dates that define seasonal shift, linear interpolation by time is allowed. The supplied energy for each season is

$$Q_{heatingseason}^{supplied} = \sum Q_{spaceheating}^{supplied} + \sum Q_{appliances}^{supplied}$$
$$+ \sum Q_{hotwater}^{supplied} + \sum Q_{external}^{supplied} \quad (2.27)$$

$$Q_{coolingseason}^{supplied} = \sum Q_{spacecooling}^{supplied} + \sum Q_{appliances}^{supplied}$$
$$+ \sum Q_{hotwater}^{supplied} + \sum Q_{external}^{supplied} \quad (2.28)$$

$$Q_{intermediateseason}^{supplied} = \sum Q_{appliances}^{supplied} + \sum Q_{hotwater}^{supplied} + \sum Q_{external}^{supplied}$$
$$(2.29)$$

DELIVERED SPACE HEATING

Summing the fuel bill data during the heating season makes estimation of the delivered space-heating energy possible.

The measured or estimated fuel quantities for other specific energies that make use of the fuel type that is used for heating are subtracted from the sum. This difference, divided by the seasonal efficiency of the heating units (default, estimated or measured value) and the conversion factor of the fuel, gives the estimation of the delivered space heating, mathematically expressed as

$$Q_{spaceheating}^{delivered} = \eta_{sh} \cdot CF (Q_{heatingseason}^{supplied\,fuel} - Q_{appliances}^{supplied\,fuel}$$
$$- Q_{hotwater}^{supplied\,fuel} - Q_{external}^{supplied\,fuel}) \qquad (2.30)$$

This equation assumes that the entities in the brackets are known. In other words, the total amount of fuel that has been supplied during this period has been assessed and the parts of the fuel that are distributed for other specific energy forms have been measured or have been estimated.

The required energy that compensates for space heat losses (transmission, ventilation and air leakage), which constitute the internal environment and the operation of the residence, corresponds to

$$Q_{spaceheating}^{required} = Q_{spaceheating}^{delivered} + \eta_{ap} \cdot CF \cdot Q_{appliances}^{supplied}$$
$$+ Q_{metabolic} + Q_{solarheat} \qquad (2.31)$$

Calculation of the space-heating requirement assumes that heat delivered from the heating units is known or estimated. Heat dissipated from appliances has to be estimated or measured. Estimations of heat delivered into the space, from occupants and solar irradiation, must have been computed. With the underlying assumption that a steady-state approach can be applied during the period of time considered, the required energy for space heating will equal the space heat losses. The space-heating requirement is used in normalization calculations.

With these values available, normalization calculations can commence; see Chapter 3.

DELIVERED SPACE COOLING

For the cooling season, the total supplied energy will be the sum of all bills for each fuel, given by

$$Q_{coolingseason}^{supplied} = \sum Q_{spacecooling}^{supplied} + \sum Q_{appliances}^{supplied}$$
$$+ \sum Q_{hotwater}^{supplied} + \sum Q_{external}^{supplied} \qquad (2.32)$$

The fuel that the cooling units use may also be distributed to generate other specific energies. With the fuel consumption known or appropriately estimated during the cooling season, the delivered space-cooling energy can be estimated, such that

$$Q_{spacecooling}^{delivered} = \eta_{sc} \cdot CF (Q_{coolingseason}^{supplied\,fuel} - Q_{appliances}^{supplied\,fuel}$$
$$- Q_{hotwater}^{supplied\,fuel} - Q_{external}^{supplied\,fuel}) \qquad (2.33)$$

Commonly, cooling units are powered with electricity, which is shared primarily with appliances (household appliances and lighting). This requires estimation of how much electricity is used for this purpose. A convenient method is to assess the base load of electrical power during the intermediate periods. Rated values from manufacturers or default values for different fabrications or types can also be used.

The required energy to obtain the indoor environment and operation of the residence, during the cooling season, corresponds to

$$Q_{coolingseason}^{required} = Q_{spacecooling}^{delivered} + Q_{hotwater}^{delivered} + Q_{solarhotwater}$$
$$+ Q_{appliances}^{delivered} + Q_{external}^{delivered} \qquad (2.34)$$

from which the required energy for space cooling corresponds to

$$Q_{spacecooling}^{required} = Q_{spacecooling}^{delivered} \qquad (2.35)$$

The implication of this equation is that a house that is not equipped with cooling units will not have a space-cooling requirement, since the occupants 'accept' the thermal conditions of the internal environment. Moreover, it also indicates that shading devices do not affect energy consumption in the absence of cooling units. However, if cooling units are present, shading devices will affect energy use, reflected within delivered space-cooling energy.

DELIVERED DAILY BASE ENERGY USE

The intermediate seasons are when neither space heating nor space cooling is applied. Bills that are from this period offer valuable data, which to a large extent is climate-independent. Assessment of specific energy uses from this period are what can be considered to be constant on a daily basis. With the assumption that the base daily consumption is the same for the heating and the cooling seasons, assessment of climate-dependent energy use is facilitated.

The delivered energy use during this period can be considered to be the sum of the delivered specific energies, in an equation expressed as

$$Q_{intermediateseasons}^{delivered} = Q_{hotwater}^{delivered} + Q_{appliances}^{delivered} + Q_{external}^{delivered} \qquad (2.36)$$

Division of $Q_{intermediateseasons}^{delivered}$ by the duration of the intermediate seasons t_{int} will give the base (daily) power load:

$$\Phi_{base}^{delivered} = \frac{Q_{intermediateseasons}^{delivered}}{t_{int}} \qquad (2.37)$$

If this power can be considered to be constant over the year, multiplication by the duration of the heating and cooling seasons, respectively, will simplify assessment of the space-cooling and space-heating energies.

Checks should be made, such as:

- Have the occupants been away from the residence for a longer period of time during the intermediate seasons than in the heating or cooling seasons?
- Are any appliances used differently during the seasons?
- Can external appliances be considered to be constant over the seasons, or are there additional external appliances used during the heating and the cooling seasons?

PROTOCOL FORM

Municipality   House Number   ID number

House address:

Estate property:

Owner/administrator:

Owner address (if other than house address):

Tenant Name:

Audit performed (yyyy-mm-dd):

Audit company:

Auditor name:

Figure 2.3(a) The protocol form (front page)

Corrections to seasonal base loads should be made depending on the answers to the above questions. This can involve adding or subtracting average daily power usages that are specific for each season.

THE PROTOCOL FORM

The form for BEP consists primarily of a front page with administrative information, followed by three tables and a list of footnotes.

The front page contains administrative information and, because this type of information varies between countries, a front page should be produced for each nation. The page must contain:

- house or estate identification
- house or estate location
- residential address
- occupant name
- owner/administrator name
- contact information (i.e. phone number)
- audit date
- audit company
- auditor name
- certification stamp and certification date.

The protocol form is shown as Figure 2.3(a, b, c).

MONITORED ENERGY PROTOCOL (MEP)

The Monitored Energy Protocol (MEP) is the more detailed protocol of the two. Although MEP takes into consideration monitored data, it also relies on bill information and earlier records. Complementary information is obtained from an audit, which is performed when the monitoring equipment is installed. The protocol is to be performed at a minimum cost, and this is one of the reasons that certain services provided within MEP are left optional. The occupant or owner of the residence has the freedom to choose which service, including the rating process itself, is to be performed.

Strategy

MEP can be considered to be a protocol that replaces BEP in the cases where the requirements of BEP are not fulfilled. MEP requires measurements. The measurements will in most cases improve the accuracy of the results. Instead of the specific energies being estimated, these will be directly or indirectly measured over a minimum period of two weeks. During this relatively short time period, the order of magnitude of specific energies will be assessed and indoor temperatures will also be measured.

The monitoring period is required to be a minimum of two weeks. However, the duration will to a large extent be dependent on what the monitored data is to be used for. If the purpose is to determine more accurate values of specific energy than in the case of BEP, two weeks may be sufficient. However, if it is desired to determine the heat-loss factor of the building by means of monitoring, the period may be extended to some 6–10 weeks or longer. If the bills can be considered accurate enough and the billing frequency sufficient, a two-week measurement of indoor temperatures may be enough to use some other method, for example the energy signature method or the UA and gA method, for estimating the heat-loss factor. This assumes that the indoor temperature during the monitoring period was the same as during the billing periods. In other words, the service required and the constraints determine the length of the monitoring period.

Measurement strategy

The measurements are to be performed for a minimum period of two weeks. The aim during this period is to assess, as part of the internal and external environments:

- the indoor temperature
- the outdoor temperature
- the degree-hours; these can directly be measured or calculated from the temperature measurements
- solar radiation and relative humidity (optional if these variables are considered to influence the energy balance of the building during the period in which the measurements are made).

For the monitoring period, the supplied energy must be recorded. This applies to both billed and non-billed supplied fuels. Depending on the availability of existing metering devices, it may be necessary to install a sensor. Another possibility is for the auditor periodically, during the monitoring phase, to take readings. This depends on what the data will be used for. In the event that monitored data is required for estimation of the heat-loss factor, readings have to be performed at least on a weekly basis.

Another assessment that is required within MEP is to determine the efficiency of energy conversion systems. This can be done by two means:

- direct measurement of the efficiency
- an approved equivalent method.

Form table 2 List of energy conversion systems, their efficiencies and how these were assessed

SUPPLIED ENERGY	Type 1	Type 2	Type 3
Estimated/Measured			
Conversion factor			
Billed: period 1			
Billed: period 2			
Billed: period 3			
. . .			
Non-billed: period			
SUM (kWh/year)			

Form table 1 Supplied energy

ENERGY CONVERSION SYSTEM	Fuel type	Winter efficiency	Summer efficiency	Estimated/Measured/Calculated
System 1	Type 1			
System 2	Type 2			
System 3	Type 3			
. . .				

Figure 2.3(b) The protocol form (Form tables 1 and 2)

Form table 3 Allocation of delivered energy, on an energy conversion system level, to each of the specific energies

Method	Footnote 1	Footnote 2	Footnote 3	Footnote 4	Footnote 5
SYSTEM 1	Space heating	Space cooling	Appliances and lighting	Tap hot water	External appliances/spaces
Period 1					
Period 2					
Period 3					
. . .					
SUM					

Method	Footnote 6	Footnote 7	Footnote 8	Footnote 9	Footnote 10
SYSTEM 2	Space heating	Space cooling	Appliances and lighting	Tap hot water	External appliances/spaces
Period 1					
Period 2					
Period 3					
. . .					
SUM					

Method	Footnote 11	Footnote 12	Footnote 13	Footnote 14	Footnote 15
SYSTEM 3	Space heating	Space cooling	Appliances and lighting	Tap hot water	External appliances/spaces
Period 1					
Period 2					
Period 3					
. . .					
SUM					
SUM (all) kWh/year					

FOOTNOTES

1. . . .
2. . . .
3. . . .

Figure 2.3(c) The protocol form (Form table 3)

Approval of equivalent methods has to be on a national basis. A possible method is, for example, proposed within the Save Belas project.[3]

A vital part of the measurements is to perform sub-metering of specific energies, unless the supplied energy is used only for one specific energy (single-use fuel). The specific energies are repeated below as a reminder:

- energy for space heating
- energy for space cooling
- energy for heating domestic water

- internal appliances and lighting
- external appliances/spaces and lighting.

The measurements of supplied or delivered energy during the minimum two-week period are intended to serve as a basis for finding the distribution of energy for multipurpose energy sources, especially to obtain reliable values for space heating and space cooling. Furthermore, measurements can be performed on non-metered energy sources, or on energy sources and fuels that are not registered in bills. MEP will strive to verify the non-billed information given by the end-user.

For climatic variables, measurements will be performed with equipment brought and installed by the auditor. Specifications on technical data (requirements), placement and sampling rates of the various gauges are listed in Chapter 2. The auditor is held responsible for the set-up and the functioning of the data-acquisition equipment.

During the period of measurement, sub-metering of supplied energy is to be performed. Because of the wide range of fuel and system types, and the various combinations of systems, it is the task of the auditor to determine how the monitoring is to be performed at a minimum cost.

Since the measurement procedures will require that certain meters and sensors be installed on site, the auditor will be required to have authorization in certain fields or be accompanied by a person with authorization. Primarily authorization is need within the electrical and HVAC fields.

PROCEDURE STRATEGY

A step-by-step illustration is given below, summing up the major steps involved in MEP. The steps are described in more detail in the sections below. They involve the following:

- **Pre-arrangements:** The auditor informs the occupants about the experimental protocol, how it works, limitations etc.
- **Collection of billed data:** Series of bills are collected.
- **Collection of non-billed data:** Series of non-billed quantities are collected (on the basis of information supplied by the end-user or a third party).
- **The audit:** General information is assessed for the building. The general information covers aspects such as:
  - technical data on the building itself
  - technical data on the energy conversion systems and the energy distribution within the residence
  - end-user behaviour.
- **Installation of necessary gauges:** During the audit, the auditor has the possibility of investigating what sensors are needed, and where these should be placed. The auditor takes readings on existing meters.
- **Extended services:** By extending the measurement period, estimation of the heat-loss factor of the residence is possible. This type of data also allows normalization calculations based on normalized indoor conditions. Moreover, miscellaneous measurements can be

performed on rates of ventilation, indoor air quality, air-tightness and thermal bridges, and U values of components in the building envelope.

- **Removal of monitoring equipment:** Monitoring equipment is removed, and the auditor takes final readings on installed/existing meters.
- **Analysis of bills, together with monitored data:** The analysis will take into consideration:
  - supplied energy or delivered energy from each energy conversion system
  - estimated annual delivered energy from each energy conversion system, based on measured entities
  - the allocation of each delivered energy to the correct specific energy, special attention being given to the specific energies for space heating and space cooling.
- **Results:** The outcome of MEP is the following:
  - an annual record of supplied energy
  - an annual record of estimated values for the five specific energy types.

    This output is directly comparable with that of BEP. Moreover, the extended services will provide informative results that explain why the energy use is the way it is.

Pros and cons

MEP consists, from the energy point of view, of a very short monitoring period. The objective of the energy measurements is primarily to assess the distribution of delivered energy among the various specific energies. Secondarily, energy use that is climate-independent is important to identify. The same applies to non-billed fuels. Moreover, internal temperatures are assessed – both temperature levels and patterns. The effect of the temperatures is the first introduced into the normalization procedures.

The values obtained from the monitoring campaign, together with billed information, contribute to an enhanced accuracy in the results when compared with BEP results. Whereas BEP relies on default-value estimations, MEP provides actual values. The underlying assumption is that, for example, the mean daily climate-independent specific energy use obtained during the monitoring period is the same throughout the season or year considered. MEP can be considered to provide BEP with complementary information by means of measurements. Nevertheless, within MEP the bills will still have to be analysed and are the essential energy data within the methodology. The only means of not relying on bills would be to deploy a heavy monitoring scheme lasting for a year or more: this would provide true annual data on energy use and temperatures.

The extended monitoring period gives more reliable values than those of a two-week measurement. In a successful monitoring campaign, correlation between parameters and variables may be found. An example is the heat-loss factor of the building. For the purpose of determining the heat-loss factor of the building, the extended time is dependent on which method is being used. Other

Table 2.5 Limitations of MEP

Aspect	Assessment	Comment
Indoor climate	Measured* Based on information from the occupant[†]	Patterns and levels are obtained from the monitoring period. These are applicable only to the type of season when the measurements were made. Set-point temperatures of the system are, if possible, assessed during the audit
External climate	Measured* Climate station data[†]	Data from the nearest climate station can be used. Discrepancies in micro- and macroclimate can be assessed during the monitoring period. For annual calculations, data from the nearest climate station should be utilized
Energy conversion system efficiencies	Measured*	These are measured, directly or implicitly, during the monitoring period. Seasonal variations make use of national default values
Sub-metering of specific energy	Measured* Annual estimation[†]	Time-consuming and costly. Specific energy is calculated for the monitoring period, but estimations have to be made for the rest of the considered (annual) period. Key values, such as daily climate-independent energy, are obtained from the monitoring period. (Estimations can be done using default values, statistics or prescribed codes. These are to be determined on a national level, to suit the framework for country-specific constraints)
Solar energy and metabolic heat	Estimated[†]	Can be calculated from relevant assessed values (solar apertures and occupant presence), but with large uncertainty. Solar energy will require computational procedures
Non-billed supplied energy	Measured* Annual estimation[†]	Non-billed supplied-energy information given by the end-user can be verified or recalculated on the basis of measured values. Analysis has to be done to consider whether or not the consumption is climate-dependent; this may lead to an extended monitoring period
Heat-loss factor	Measured using environmental temperatures and the heat delivered over a period	Usually requires an extended measurement period. If the billing frequency and quality are adequate, and corresponding temperatures are known for these periods, the heat-loss factor may be determined without the requirement of the extended period

* Measured within MEP.

[†] Estimated on an annual basis.

influencing factors are the building type, the energy systems and the prevailing internal and external climates.

The limitations of MEP are listed in Table 2.5, which includes some comments on the reasons for the limitations.

Conditions for application of MEP

Certain requirements have to be fulfilled for the MEP procedure to be applied. These primarily concern the quality of the series of bills, which within MEP are mandatory data.

THE REQUIREMENTS CONCERNING THE BILLED AND RECORDED ENERGY CONSUMPTION

- A complete series of bills or readings from gauges that correspond to the annual total energy consumption is required. The bills must at least contain information on the fuel type, the quantity of fuel and the time of delivery, as well as whether or not the billed values are predictions or measured values.
- Conversion factors from fuel types to the unit Wh are necessary. The specific energy content of each fuel type is determined at a national level.

- The efficiencies of energy conversion systems that could not be measured have to be obtained by using default values from national codes or by using other procedures.

THE REQUIREMENTS CONCERNING AUDIT AND CLIMATE INFORMATION

- Reliable climate data must be available from the closest climate station, and for the two monitoring weeks must also be measured on site. Reference climate data must exist for all regions.
- The energy auditor must have inspected the site, accessed all premises, set up and read all relevant gauges, measured the relevant variables and accurately filled in the audit form. It is presumed that the resident representative provides reliable information on behaviour and utilization patterns.
- A criterion that must be fulfilled is that the expected global energy consumption during the two weeks of the monitoring period should correspond to at least 5% of the annual global energy consumption.
- When gauges cannot be installed for practical reasons, default values will have to be used (in the same manner as for BEP). The limitation on this is that the variable in question must not influence the heat balance of the building by more than 10%. Otherwise, rating is not possible. The default values should be estimated on the basis of reliable codes (determined at a national level) or using alternative methods based on calculations or statistics.

Specifications and requirements

Monitoring of climate variables will require a sampling frequency of half an hour. Detailed information is given in the sections below.

For measurements of energy flow (global and sub-metering), the performance is dependent on the meter and gauge types that are actually present, and also on the possibility of installing new ones. The requirements are set as follows:

- For electronic monitoring equipment, the sampling rate should be hourly.
- Older monitoring equipment (already on site) should be read at the beginning of the monitoring period and at the end of the monitoring period. However, electronic monitoring, with shorter sampling intervals, is preferred. The occupants may perform readings after being instructed by the auditor on how to do this.

The inaccuracy of the meters or complementary equipment/methods should be no more than ±5%.

The choice of monitoring period during the year should be during one of the seasons when the energy requirement is relatively large. *A criterion that must be fulfilled is that the expected global energy consumption during the two weeks of the monitoring period should correspond to at least 5% of the annual global energy consumption.* Judging which two consecutive weeks of

Table 2.6 Room preference for location of temperature gauges, starting with the highest priority[4]

Gauge 1	Gauge 2
Living room	Hall
Hall	Bedroom
Bedroom	Kitchen
Kitchen	

the year are most suitable can be done on the basis of old energy bills.

From the point of view of the extended monitoring period, which primarily aims at determining the heat-loss factor, the requirements of the method to be used should be followed. For example, the energy signature method gives more reliable results when the external temperature and the solar radiation vary significantly during the monitoring period. In other methods, better results may be obtained if climate conditions are as constant as possible.

CLIMATE GAUGE SPECIFICATION

The following information on environments has to be collected for a period of time:

- indoor temperature
- outdoor temperature
- solar radiation and outdoor relative humidity (optional).

The placement and technical specifications of gauges are described below for the various types of measurements.

Indoor temperatures

The chosen location for the gauge is one where the temperature is representative of the thermal zone/room. The room should be one that is frequently used, the doors to the room should normally be open, and the heating/cooling devices should be set to 'normal' room temperature.

The choice of room, or temperature zone, is to be chosen according to the ranking specified in Table 2.6. The ranking is highest in the first row, with descending preference for lower rows.

Some buildings may require more than one temperature gauge. This depends on the number of thermal zones present in the building. Volumes within a building that have a temperature difference greater than 2°C should be considered as different thermal zones.

Indoor temperature gauges are *not* allowed to be placed

- on external walls
- closer to external construction joints than 0.5 m
- closer to external doors or windows than 3 m
- on internal walls that on the other side have chimneys or air ducts
- on internal walls that on the other side have refrigerators or freezers
- on walls that for more than half the day are exposed to direct sunshine

- on walls above radiators or other heat sources
- on walls above televisions, PCs or other electrical appliances
- closer to lamps than 0.5 m
- closer to fireplaces or furnaces than 1 m
- so that the gauges are exposed to mechanical stress
- so that the supply air is directed towards the gauge.

Technical specifications

The technical specifications in terms of degree-hours, accuracy and temperature, and sampling rate and data collection are given in Tables 2.7, 2.8 and 2.9.

Outdoor temperature

Only one outdoor temperature-measuring device is needed. For this reason, the location has to be carefully chosen. The outdoor temperature gauge is *not* to be placed so that:

- direct sunlight influences the gauge; guages are best mounted on north façades, preferably sheltered from precipitation by roof eaves
- exhaust air is directed towards the gauge
- external doors and open windows are closer than 1 m.

Technical specifications

The technical specifications in terms of degree-hours, accuracy and temperature, and sampling rate and data collection are given in Tables 2.10, 2.11 and 2.12.

Solar radiation (optional for local measurements)

Values can be collected from the nearest climate station, if it is considered that building energy gains due to solar radiation are small compared to the heat loss of the building.

In climates with strong solar irradiation, the solar radiation may be measured. The sensor (pyranometer) should at least fulfil the requirements of ISO 9060 Second Class.[5] For such measurements, the sensor should be placed

Table 2.7 Ranges for degree-hours – Indoor temperature

Degree-hours	$\pm 9{,}999°C \cdot h$
Temperatures	At least 0 to $+50°C$

Table 2.8 Accuracy, temperature – Indoor temperature

Calibration at	$20°C$
Absolute inaccuracy	Less than $\pm 0.5°C$
Discrepancy between indoor and outdoor gauges	Less than $0.2°C$
Discrepancy between indoor gauges	Less than $0.1°C$
Long time instability (-30 to $+30°C$)	Less than $\pm 0.5°C$
Thermal time constant	Around 2 s
Non-linearity (-30 to $+30°C$)	Less than $\pm 0.2°C$

Table 2.9 Sampling rate, data collection – Indoor temperature

Sampling rate	0.5 h
Data collection	Daily or hourly mean of sampled values

Table 2.10 Ranges of degree-hours – Outdoor temperature

Degree-hours	$\pm 9{,}999°C \cdot h$
Temperatures	At least -28 to $+50°C$

Table 2.11 Accuracy, temperature – Outdoor temperature

Calibration at	$0°C$
Absolute inaccuracy	Less than $\pm 0.5°C$
Discrepancy between indoor and outdoor gauges	Less than $0.2°C$
Discrepancy between outdoor gauges	Less than $0.1°C$
Long time instability (-30 to $+30°C$)	Less than $\pm 0.5°C$
Thermal time constant	Around 2 min
Non-linearity (-30 to $+30°C$)	Less than $\pm 0.2°C$

Table 2.12 Sampling rate, data collection – Outdoor temperature

Sampling rate	0.5 h
Data collection	At least a daily mean value of sampled values

Table 2.13 Accuracy – Solar radiation

Nominal spectrum	$0.3–3.0\,\mu m$
Non-linearity ($<1{,}000$ W/m^2)	$\pm 3\%$
Temperature range	At least -30 to $+60°C$

horizontally for global solar radiation assessment. The site must not be shaded.

The accuracy specification is given in Table 2.13.

Humidity sensor (optional)

Humidity gauges should be placed close to the outdoor temperature gauge and fulfil the same placement requirements as the outdoor temperature gauge. This type of measurement is only of interest if the monitoring campaign includes energy for space cooling in a humid climate. This is because moisture has an impact on the performance of the cooling units.

MEASUREMENT OF SUPPLIED ENERGY

Because of the vast number of types of systems that exist in the building stock, it is virtually impossible to prescribe how the measurement is to be done for all cases. The auditor is held responsible for the measurements being carried out properly and for their being relevant to the results obtained.

The important issues that must be assessed by the measurement of the supplied energy for a minimum of *two weeks* are:

- the total supplied energy during this period
- the distribution of the supplied energy in terms of the specific energies
- the non-billed energy supplied.

The gauges and meters that already exist in the building should be used to the greatest extent possible, after it has been checked that the meters work. The setting up of meters should be performed so as to minimize the destruction or modification of the systems. In cases where metering is not possible, for practical or economical reasons, the assessment methods of BEP can be used. This must be carefully noted on the completed forms.

With electronic monitoring equipment, the sampling rate should be every half an hour. In some cases, the rate may be more intensive, down to seconds. It may not be possible to equip already existing meters with electronic data acquisition. For this reason, these types of meters must be read by the auditor at the beginning and at the end of the monitoring period. Residents may read meters manually during the period, but will not be compelled to do this. Readings made by residents will require instructions from the auditor.

Use of electricity

Most residential premises are equipped with at least one electricity meter on site, which may be used. Many brands have built-in possibilities for performing pulse counts. This may require an energized port as part of the monitoring equipment. These types of meters commonly monitor the total electricity supplied to the building.

In the case where there is a meter that does not provide the possibility of counting pulses, the readings may have to be made manually by the resident or may require complementary equipment. It is desirable that the auditor interferes as little as possible with the electrical power supply of the building, as this requires special authorization.

The central electrical fuseboard (consumer unit) provides the possibility of monitoring branch or distribution circuits. *Inductive gauges* can be used in this case and mounted on the branch circuits for space-heating systems or devices, space-cooling systems or devices and/or domestic water heating. The electrical consumption for other purposes is assumed to be for household appliances and lighting.

For devices that are part of a specific energy other than household appliances and lighting, and are powered by means of socket outlets on the wall, *socket mounted energy meters* can be used. These are convenient if, for example, portable electrical radiators are used.

Technical specification

The inaccuracy of the meter or the complementary equipment should be no more than ±2%.

Use of oil

Determination of oil consumption during the monitoring period can make use of oil flow meters that already are installed in the system. However, the performance of the oil flow meter should be checked with either *the nozzle* or *the cup method*, which are described below. If no oil flow meter exists, one of the two methods below should be used to investigate the oil flow through the burner. When this is done, a *running-time sensor* must be used. Fuel consumption during the measurement periods is the run time multiplied by the flow.

The other option is to install a flow meter, which should be of a type that is compatible with the data acquisition equipment.

Oil burners are commonly marked with the following information:

- manufacturer and model type
- oil flow rate in USgal/h and sometimes litres/h (litres/h = 3.785 · USgal/h).

These flows are based on a pump pressure of 7 bar, an oil density of 0.82 kg/litre, an oil viscosity of 3.4 cSt (mm^2/s) and a temperature of 20°C. However, the stated flows may not be accurate. According to manufacturers, the flow may vary by ±5%.

Some systems may be equipped with oil flow meters. These could be used, but may have inaccuracies of up to ±10%. The inaccuracies may be greater if contaminants (such as sand) in the oil disturb the function of the flow meter. The oil flow meter can be used as a gauge if the inaccuracy is less than ±5%.

For these purposes, two methods, the nozzle and graded cup method, are proposed for determination of oil consumption during the measurement period. The methods require that the run time of the burner be recorded during the monitoring period.

The nozzle method

The nozzle method is the easier one to use. The nozzle is dismounted so that it can be instrumented with a manometer to measure the pressure of the pump. A correction factor is used to multiply the marked oil flow rate of the burner (at standard conditions as stated above), where this factor is defined as

$$K = \sqrt{\frac{measured\ pressure}{7}}$$

The required time is about 30 min and the estimated inaccuracy around ±5%.[6]

The graded cup method

The graded cup method requires that the oil burner be dismantled. Between the nozzle and the burner body, an extension hose (produced for this purpose) is connected so that the nozzle can be directed into a graded cup. For a recorded period (for example, 30 minutes) the amount of discharged oil is measured.

The required time is about 1 h and the estimated accuracy around ±2.5%.[6]

Comments

The configuration of boilers and burners can vary considerably, which requires certain technical skills on the part of the auditor. However, a specialist from the manufacturer is not needed if the technical instructions on the equipment and procedure are well produced.

When this measurement has been made, it is recommended that there should be no new oil deliveries during the monitoring period. One reason for this is that the viscosity of oil can vary between 3.0 and 4.5 cSt. A change in viscosity of 1 cSt can lead to a change in flow corresponding to 2–6%. Another reason is that the heat content of oil may vary by 0.75% between different deliveries.

In the case where sensors cannot be installed, readings may have to be performed on differences in the oil level within the tank.

Technical specification

The inaccuracy of the meters and the complementary equipment/methods should be no more than ±5%.

Use of natural gas

Gas is often transported in permanent pipes, which are equipped with meters (for billing global consumption). These meters can either be complemented with a monitoring device or be read manually.

It is unusual for sub-metering to be performed in the branches of the piping system. This proves to be a problem if gas is used for space heating and tap-water heating (a common combination) and for cooking. For sub-metering, see the section above on *Tap hot-water measurements*.

When liquefied petroleum (bottled gas) is used, the resident is asked to give information on how the bottles are used, i.e. for heating or cooking. The simplest means of measuring liquefied petroleum use is to weigh the bottles that are used during the monitoring period. Another option is to fit a measurement device to supply tubes.

Technical specification

The inaccuracy of the meters and complementary equipment/methods should be no more than ±5%.

Biomass/wood fuel, coke/coal

Fuel in the form of biomass, wood, coke or coal is conveniently weighed or measured (by volume) prior to and after consumption. The auditor should perform weighing and/or volumetric measurements at the start and at the end of the monitoring period. Residents/managers may perform daily measurements during the period, but this is not mandatory.

District heating/cooling

District heating and cooling is often equipped with energy meters. These are generally based on a form of flow measurement and a temperature-drop measurement. Should there be no energy meters present, such a meter will have to be installed.

Technical specification

The inaccuracy of the energy metering should be no more than ±5%.

Tap hot-water measurements

The energy source for heating domestic water is usually the same as, or integrated with, that for space heating (electricity, gas and oil). Energy for the heating of domestic water is commonly not measured in conventional systems in buildings. The aim of the measurements will therefore to a great extent be to determine the part of the fuel consumption that is used for space heating and the part that is used for heating domestic water.

Energy for hot water is to be assessed by measuring one of the following:

- the flow of hot water (hot-water consumption) and the temperatures
- the heat supplied to the water.

In the first case, a water flow meter has to be installed in the pipe going to the end-user. The energy used can be estimated to be the amount of hot water that has been used times the rise in temperature, the density and the specific heat capacity. This corresponds to the temperature that the water has on entering the boiler and the temperature in the accumulation tank, especially if the water has been pre-heated by some device.

If monitoring possibilities are limited, an approximation is to assume that the temperature of the incoming water is the mean annual outdoor temperature, and that the tank temperature is the same as the set-point temperature. This assumes that the domestic water has not been pre-heated.

In the second case, the heat delivered to the water is easily measured if electrical heating is used. Running-time sensors or inductive gauges that are connected to the heating coils of the system can be used for this. The power of the heating coils is often available.

Technical specification

The inaccuracy of the energy metering should be no more than ±10%.

Comments

In some cases heating of water taking place in the building is not worthwhile measuring. An example is dish- and clothes washers that heat water in the washer itself during utilization. The energy source here is electricity and is therefore allocated to the household appliances. However, other models use water that has already been heated and, in this case, the energy use will be a part of the domestic water heating.

Processing of data

The results obtained from the monitoring campaign give indications of the magnitude of the specific energies. Distribution of energy from each source (fuel type) to a specific energy has been assessed during the monitoring period. In the cases where BEP values were based on estimations, MEP produces a better estimation on an annual basis.

The description presented below, on how to process data, depends on what purpose the data serves. Two types of monitoring schemes can be used:

- a relatively short monitoring period, in the region of two weeks
- a longer monitoring period of around 6–10 weeks.

THE SHORT MONITORING PERIOD

The short monitoring period, somewhat longer than or equal to two weeks, will give limited but yet important information from an energy perspective. Provided that the occupants have lived in a 'normal' way, the monitored data will give an indication of:

- the size of delivered energies
- the size of and the interrelation between specific energies
- the internal climate (temperature and utilization patterns of the energy conversion systems)
- local external climate variables, as opposed to the corresponding values from the nearest climate station.

For the purpose of assessing values that are to be used for rating the building, information from the short monitoring period can be treated in the same fashion in BEP. For this to work, it is necessary to make the following assumptions concerning the acquired data:

- Mean base loads measured during the two weeks are constant throughout the season during which measurements were conducted.
- The ratio between the energy for tap-water heating and the water consumption during the period is almost constant during the considered season.
- Measured temperature patterns are similar, throughout the entire season, to those that were measured.
- Non-billed fuel consumption is summed for the period. This quantity is divided by the mean temperature difference between the monitored internal and external climates, and is divided by the number of monitored days.

Now, billed information will be used in the same way as in BEP, with the following steps:

- First, the length of the considered season is defined and the corresponding bills are compiled in order.
- The base loads, multiplied by the duration of each billed period, will give the base energy and the delivered energy that this requires. The delivered energy divided by the system efficiency gives the fuel quantity that is used for this purpose.
- Base-load fuel quantities are subtracted from the billed quantities. The remainder is climate-dependent: fuel that was used for space heating or space cooling.
- Non-billed fuel consumption, which is considered to be climate-dependent, is calculated for the corresponding billed period. Based on the assumption that fuel consumption ceases (= 0) at the end of the season (the temperature difference between the internal and external environment is known), linear interpolation with the temperature difference as abscissa is used to estimate the annual non-billed fuel consumption.

From this point, BEP procedures can be followed and the normalization and rating processes can commence.

The forms used for BEP are also used for recording the information. Therefore, as in BEP, one of the forms includes the data, and there is also an equivalent form but including measured values. The footnotes to the first form indicate what values are used from the measured series.

A complementary calculation can be carried out if the quality and frequency of bills is such that these, together with monitored energy use and indoor temperatures, can make use of the energy signature method (see below). The underlying crucial assumption is that the temperature that was monitored during the short-term period is the same throughout the considered season. With billed data and climate variables, the use of the energy signature method or a corresponding method can be used to assess the heat-loss factor of the residence.

THE EXTENDED MONITORING PERIOD

The monitoring period must be longer than two weeks if the criterion of 5% energy use is not fulfilled. However, the main motivation for extending the monitoring period is to determine the quality of the building envelope, which is reflected in the heat-loss factor.

There are several ways of assessing the heat-loss factor, but the costs for these can vary considerably and some methods require the occupants to move out of the residence. The aim of MEP is to assess the heat-loss factor while the occupants continue their normal lives during the monitoring period.

Estimation of the heat-loss factor

The method used for assessing the heat-loss factor should be determined at a national level. The motivation is the following:

- The ways in which building technologies and systems are used will give different monitoring periods.
- The external climate will influence the length of monitoring periods and determine which parameters should be considered in the fundamental equations.
- It is preferable to use traditional practices and national norms, since the ways methods are used and the accuracy of the results is to some extent dependent on experience.

Below, the energy signature method is shown to illustrate how such a method may be used.

The energy signature method

The energy signature method (ES method) is commonly used to estimate energy consumption and can be used to determine the heat-loss factor of a building. Based on energy measurements, the model extrapolates building energy performance over a longer period than that during which the measurements took place. The method can also be used to normalize energy consumption.

Within the framework of this methodology, the ES method will not use the monitored data from the two weeks. The length of the monitoring period is too short for the ES method to give reliable results. Here, the strategy is, during the two weeks, to assess the *mean indoor temperature* of the building. This temperature is then assumed to be constant throughout the season. By using information from the energy bills and accessing climate data for the billed period, the ES method can be applied.

Description of the model

The measured energy consumption is statistically regressed against climate data. The model is formulated mathematically for an individual building unit[7] such that

$$E^{supplied} = (c + b \cdot \Delta\theta)T + f \cdot \bar{I} + d \cdot P \text{ (Wh)} \qquad (2.38)$$

where:
E = the annual total energy use (Wh)
T = the length of the heating season (h)
P = the length of the non-heating season (h)
c = a temperature-independent rate of energy consumption (Wh/h) during the heating season
d = a temperature-independent rate of energy consumption (Wh/h) during the non-heating season
b = the gross heat-loss factor of the building unit (W/K)
f = the solar aperture of windows (m²)
$\Delta\theta$ = the mean sampled temperature difference between internal and external environments (K)
$\bar{I}$ = the average sampled solar radiation (Wh/m²)

The terms c, b, f and d are called *energy parameters* and have specific values for each building.

Physical interpretation of the energy parameters

The ES method is in essence a black-box approach, giving limited information on how energy is consumed. However, vital information can be obtained if the energy parameters are interpreted.

The gross heat-loss factor of the building unit is

$$b = \frac{\sum_{i=1}^{n} (U_i \cdot A_i + \psi_i \cdot \ell_i) + 0.33(1 - \eta_{vent}) \cdot n \cdot V}{\eta} \text{ (W/K)} \qquad (2.39)$$

where
$\sum_{i=1}^{n} (U_i \cdot A_i + \psi_i \cdot \ell_i)$ = is the heat-loss factor of the building envelope, including effect of thermal bridges (W/K)
n = the average rate of air changes (1/h)
V = the enclosed volume of the building unit (m³)
η = the efficiency of the heating/cooling systems (dimensionless)
η_{vent} = the efficiency of air-handling systems (dimensionless)

The winter factor c is climate-independent and primarily related to the number of occupants and their behaviour:

$$c = \frac{E_{hot\ water} - E_{persons} - E_{household}}{\eta \cdot T} \qquad (2.40)$$

where
$E_{hot\ water}$ = the energy required for heating domestic hot water
$E_{persons}$ = the metabolic energy from occupants
$E_{household}$ = the electricity for household appliances

The summer factor d has the same formulation as Equation 2.40, compensated for the length of the summer season.

The solar aperture f is a parameter that takes into account solar irradiation within the energy balance of a building. Moreover, the orientation of the house has to be taken into consideration. f is given by

$$f = \frac{\sum_{d=1}^{4} (1 - s_d) \cdot T_d \cdot A_d \cdot I_d}{\eta \cdot I_{ann}} \qquad (2.41)$$

where
s_d = the shading factor of the external environment (dimensionless)
T_d = the transmission factor of the glazed part of the windows (dimensionless)
A_d = the glazed area of the window (m²)
I_d = the solar radiation on the window surface (Wh/m²)
I_{ann} = the average global radiation on a horizontal surface at the geographical location of the building (Wh)

The gathering and processing of data

Given the solar radiation I, the length of the heating and non-heating seasons T and P, and the number of degree-hours $\Delta\theta \cdot T$ from climate files (from the nearest climate station), the energy parameters can be determined by regressing energy consumption with outdoor climate. Energy consumption is obtained from the bills. By means of interpolation, billed values that are obtained during different time periods are synchronized.

The part of Equation 2.38 that is climate dependent involves the first two terms on the right-hand side of the expression. By setting the $d \cdot P$ term aside and dividing the remainder of the expression by the length of the heating season, T, a new formulation is obtained which gives the average energy use per unit time w (kWh/h) during the heating or the cooling season. An error term ε is also added such that

$$W = c + b \cdot \Delta\theta + f \cdot s + \varepsilon \qquad (2.42)$$

This is a *static model*, which is valid provided that the observation period is long enough for the heat stored in or released from the building to be negligible compared to the total energy supplied during the period. With the observed data on energy consumption W, the temperature difference $\Delta\theta$ and the solar irradiation s, the parameters c, b and f are estimated with *linear regression*. If the error value ε exceeds what is acceptable, more complicated models may have to be applied; see below.

Application of the static model implies that the observed data are based on average or cumulated measured values that correspond to those of Table 2.14.

Where permanent gauges exist and continuous communication and data transfer are possible, hourly

Table 2.14 Observed values for the model

Energy signature method	Average values of $\Delta\theta$ and s	Monitoring periods
	Cumulated values of w	
Static model	Weekly, monthly	At least 10 observations

average and cumulated measured values can be used as input to the ES method. This requires a dynamic model and introduces time t.

The energy use $W(t)$ during hour t is assumed to be a linear function of variables at times t, $(t-1)$ and $(t-2)$. A simple dynamic ARX (AutoRegressive model with eXogeneous inputs) model is expressed as

$$W(t) + a \cdot W(t-1) = c_1 + b_1 \cdot \theta_{in}(t)$$

$$+ b_2 \cdot \theta_{in}(t-1) + b_3 \cdot \theta_{out}(t) + b_4 \cdot \theta_{out}(t-1)$$

$$+ f_1 \cdot s(t) + f_2 \cdot s(t-1) + \varepsilon(t) \tag{2.43}$$

With steady-state conditions, the thermal parameters are estimated such that

$$c = c_1/(1+a)$$

$$b = (b_1 + b_2)/(1+a)$$

$$f = (f_1 + f_2)/(1+a)$$

The parameters are estimated by minimizing the sum of squares of the error term $\varepsilon(t)$.

Equations for estimating the normalized energy consumption

This section is directly taken from Westergren *et al.*[8] and describes the procedure for estimating the normalized energy consumption.

Assume that data are available for k time periods ($i = 1, 2, \ldots, k$), for which average measurements have been obtained, and consider the following regression model with p explanatory variables and a constant term:

$$W_i = c + b \cdot \Delta\theta_i + f \cdot s_i + \cdots + \varepsilon_i \tag{2.44}$$

or in matrix form

$$\mathbf{W} = \mathbf{X} \cdot \boldsymbol{\beta} + \boldsymbol{\varepsilon} \tag{2.45}$$

To estimate the total heating energy use during a normalized year, a method suggested by Salkever[9] is used, whereby the least-squares estimates are obtained together with the predictions for a normalized year.

Let $\mathbf{z}^T = [1 \; \Delta\bar{\theta} \; \bar{s} \; \ldots]$ be a vector of the averages of the p explanatory variables for a normalized year. Then expand the regression models (Equation 2.45):

$$\begin{bmatrix} \mathbf{W} \\ 0 \end{bmatrix} = \begin{bmatrix} \mathbf{X} & 0 \\ \mathbf{z}^T & -1 \end{bmatrix} \begin{bmatrix} \boldsymbol{\beta} \\ \boldsymbol{\gamma} \end{bmatrix} + \begin{bmatrix} \boldsymbol{\varepsilon} \\ 0 \end{bmatrix} \tag{2.46}$$

By using the partitioned inverse formula for matrixes, the Ordinary Least Squares (OLS) estimate of the parameter vector

$$\begin{bmatrix} \boldsymbol{\beta} \\ \boldsymbol{\gamma} \end{bmatrix} \tag{2.47}$$

is then obtained as

$$\begin{bmatrix} \boldsymbol{\beta} \\ \boldsymbol{\gamma} \end{bmatrix} = \begin{bmatrix} (\mathbf{X}^T\mathbf{X})^{-1}\mathbf{X}^T\mathbf{W} \\ \mathbf{z}^T(\mathbf{X}^T\mathbf{X})^{-1}\mathbf{X}^T\mathbf{W} \end{bmatrix} \tag{2.48}$$

where $\boldsymbol{\beta}$ is the original OLS-coefficient vector and $\boldsymbol{\gamma}$ is the predicted energy use per hour for $\mathbf{z}$, a normalized year.

An expression for the variance of the prediction is obtained by noting that the total variance for the expanded model (Equation 2.48), σ_{tot}^2, is

$$\sigma_{tot}^2 = \frac{\begin{bmatrix} \boldsymbol{\varepsilon} \\ 0 \end{bmatrix}^T \begin{bmatrix} \boldsymbol{\varepsilon} \\ 0 \end{bmatrix}}{k+1-(p+1)} = \frac{\boldsymbol{\varepsilon}^T\boldsymbol{\varepsilon}}{k-p} = \sigma^2 \tag{2.49}$$

which equals the total variance in the unexpanded model (Equation 2.45). Thus the estimated standard error (se) of the estimate $\mathbf{g}$ is

$$se(g) = \hat{\sigma}(1 + \mathbf{z}^T(\mathbf{X}^T\mathbf{X})^{-1}\mathbf{z})^{0.5} \tag{2.50}$$

with which a confidence interval is obtained as

$$\mathbf{g} \pm t(k-p-1)_{\frac{\alpha}{2}} se(\mathbf{g}) \tag{2.51}$$

The width of the confidence interval (Equation 2.51) depends on the number of observations and on the variability of the data during the observation period. To increase the accuracy, the measurement periods should be chosen in such a way that the temperature differences and the solar irradiation vary as much as possible and the average temperature difference and the average solar irradiation are as equal as possible to those of the normalized year.

THE AUDITING PHASE

The auditing phase is primarily a complementary tool to the processing of the collected billed information. It serves to gather descriptive information on residential buildings and building performance and to reveal the living patterns and behaviour of the occupants.

Development of the audit form

The audit form is adapted from the ELIB project.[4] The ELIB project involved inspection of 1,200 residential units across Sweden, which was performed by 18 auditors during a planned time span of 6 months, with an extension of 6 weeks. The task was intended, as well as determining the indoor climate of the Swedish housing stock, to assess information on the technical and health aspects and the thermal performance of this stock. It also dealt with assessing the potential of making the buildings more energy efficient and less dependent on electricity. The ELIB protocol has therefore been extensively tested and there are statistics on the frequency of answers to each question.

The ELIB auditors took two days to learn the protocol and how gauges were to be placed. A test run of 90 houses was used to 'get the protocol right' and to 'calibrate' the staff. Within the first six weeks of the project, the auditors had to produce an audit report using a proper and acceptable protocol if they were to continue the task.

An ELIB audit took around eight hours for multi-family buildings and four hours for single-family buildings. The inspection comprised:

- a review of the plans of the building
- inspection at the site
- filling in the audit protocol
- setting up the measurement equipment.

This was for an 'average' inspection unit, where plans were available, that is in 70% of the cases.[10] The cost was around 4,500 SEK in the early 1990s (Tolstoy, 2000, private communications). One of the largest problems encountered was getting in contact with representatives of a household, agreeing upon suitable day/days for the audit and measurements of buildings that lacked plans. This took more time than was taken into account during the planning stages of the project. The eventual cost was higher because of this time 'inflation'.

Another audit form that was reviewed was the Residential Energy Consumption Surveys (RECS) form. RECS 'provides information on the use of energy in residential housing units in the United States. This information includes the physical characteristics of the housing units, the appliances utilized including space heating and cooling equipment, demographic characteristics of the household, the types of fuels used, and other information that relates to energy use. The RECS also provides energy consumption and expenditures data for natural gas, electricity, fuel oil, liquefied petroleum gas (LPG), and kerosene.'[11]

RECS 1997 involved interviews and collection of data for some 5,000 residences. 220 interviewers were trained for three days. Most interviews were conducted over the telephone. The questionnaire took on average 29 minutes to fill in and the span of 15–45 minutes covered 85% of the interviews. An electronic version of the questionnaire allowed immediate default value calculations. RECS 1997 is fully documented by the Energy Information Administration.[11]

The audit form that was developed and proposed in this book has been designed to do the following:

- take into consideration that the auditor is on site
- bring out the living patterns and behaviour of the occupants
- gather billed and non-billed fuel quantities
- collect technical information on the building and the energy conversion systems
- be used for both BEP and MEP
- take around an hour to complete.

Recommendations for audits

Important points to bear in mind during field surveys:[12]

- Do not rely on reported data and drawings before the auditor has seen the building with his/her own eyes.

- The investigation routine should be the same for the whole survey.
- Train the personnel who are to carry out the field studies, to ensure uniformity of terminology and usage.
- Carry out a pilot survey, to test the inspection report forms and routines.
- 'Calibrate' the auditors. This is very important in the case of visual inspections with few instrumental readings and many subjective assessments.
- The inspections should be carried out within a short period of time and by a small number of inspectors.

Auditing steps

- The auditor agrees a time to contact the customers (occupants/owner) and together with them inspects the building.
- The auditor informs the occupants/owners about the BEP and/or MEP procedures. This includes information on the requirements, the limitations, the various services and the inconveniences that may arise during the monitoring phase.
- The work of the audit starts with investigation of plans and technical documents. This information may be available from the local authority or from the building owner. Copies of this information are required. The information is filled in on the audit form. If the building is composed of parts that differ significantly in terms of building year or building technology, or if common surfaces (internal walls) have the same insulation level as envelope components, the various parts can be considered to be individual buildings. Semi-detached houses are considered to be detached buildings.
- Together with the building owner (or representative), an audit is performed on site. The representative is preferably a person with knowledge of heating, cooling, ventilation, building technology and electricity. If the representative is not appropriate for the task, a more suitable person may be asked for (this may be tested by means of simple questions on, for example, the heating system). The person is required to have keys to fan rooms, laundries etc.
- The audit should be conducted so as to check that the plans and interviews agree.
- If the representative cannot give information during the audit, this information should be provided as soon as possible after this, for example in a telephone conversation.
- During the audit, if there is sufficient illumination, a colour photograph of the exterior should be taken.
- In residences (apartments) subjected to MEP, an occupant must always be present because the gauges will be placed and precautions described. If the occupants are to read the gauges, the auditor must provide instructions.

The audit form

The protocol is a type of guide for collecting information (a copy is included as Appendix 1). In the future, it may be convenient to write a computer program that stores and processes the information. A benefit of doing this would be that the rating outcome (and certificate) could be produced on the occasion of the audit (BEP) or when the monitoring campaign is terminated (MEP).

Some points may have to be modified to suit country-specific properties. For example, revision of the first page of the audit form will be necessary. Data lists and that context in which each building and the site is identified will have to be adapted to national nomenclature and definitions. The administrative information requirements are as follows:

- identification of the building, site and address
- building owner, resident representative and maintenance operator (may be the same or different people)
- date of audit
- auditor name and company.

The present audit form was applied in the four test countries. After this the original form was extended to include questions on glazed surfaces.

General building information

General building information is to some extent informative, but also contains data that may be required in the rating methodology. These items of information are as follows:

A2   Building year. This parameter is important with respect to national building codes. Primarily, the building year will give an indication of how well the building is insulated.

A3   Year of major retrofit/conversion/extension. If the building has been subjected to retrofitting actions (such as additional insulation) or it has been extended, assumptions on the basis of the building year will no longer be valid. Furthermore, the thermal performance of thermal zones may be different for the various spaces.

A4   Building location. The location of the building has relevance to the climate data that should be applied and the influence of shading obstacles. A code for the closest climate station can be given.

A5   The type of building that is being audited.

A6   How many storeys there are in the building, and the number of storeys that are heated and cooled. This applies to both single- and multi-family houses. For multi-family houses, this information is important if the building has central heating and cooling without sub-metering.

A7   How the building is utilized. This is important with regard to energy consumption patterns. Energy use for non-residential purposes is to be excluded.

A8   Lengths, areas and volumes. The most important are the last two, the residential heated floor area and the total volume, because the various energy consumptions are given as areal or volumetric entities. This information is taken from the plans if possible, but has to be verified by the auditor by means of a visual survey.

A9   The distribution of apartments in a multi-family building. The main reason for this is that a multi-family building can participate in the rating procedure as a whole. This is the case when an apartment is to be rated, but there is no sub-metering of the energy consumption, the energy being distributed from a central heating system in a way that does not allow for individual measurements.

A10   The main ventilation system type. This is merely informative, although within MEP it is useful for describing the heat-loss factor of the residence.

A11   Records from the most recent tuning of the main ventilation system, which, like the information in A10, is only informative.

A12   For apartments in a multi-family building only: on which storey the apartment is situated. This gives qualitative information on outdoor climate exposure.

A13   Qualitative information on outdoor climate exposure for multi-family and varieties of multi-family buildings (see A5).

A14   Number of rooms and bathrooms. This information may be needed for BEP default value calculations. The number of rooms excludes halls and garages.

A15   Tables for listing window features (see Appendix 2). The number of glass panes (i.e. single-glazed, double-glazed or triple-glazed window) is to be filled in along with an estimation of the g factor and the curtain factor. The horizontal angle for each lateral section of each window is included. These angles are used to predict the partial shading factors for external obstacles. Finally, the total glazed area for the different windows or a representative window is given.

Residents and behaviour

Residents and their behaviour have a large influence on energy use. The main point of this section is, therefore, to obtain data on the patterns that the residents have. Much of the information has to be obtained from the resident representative.

B1   The number of people in the apartment and their ages.

B2   The presence of residents in the apartment/single-family unit during the summer and the winter seasons. This is to take into consideration vacations when nobody uses the building.

B3   The presence of residents during working days and weekends is estimated. This information is used for metabolic heat calculations.

B4   The representative is asked for an estimation of the indoor temperature during certain periods of the day (winter season). The estimate should be made on the basis of the living space that is occupied most frequently, such as the living room. Furthermore, if the heating system is shut off during certain periods, this should be noted. These details, though they may be erroneous, are valuable to determine the magnitude of indoor temperature.

B5 If an automatic setback temperature is applied (commonly for central heating units), this should be noted. It will influence the mean indoor temperature.

B6 An estimation is made by the resident representative of the indoor temperature in the summer season. Again, as in B4, this applies to frequently occupied rooms (living rooms).

B7 If an automatic setback temperature for cooling is applied, it should be noted. This will influence the mean indoor temperature.

B8 The length of the heating and the summer seasons may be obtained directly from the audit. If not, default values will have to be used for the region, as prescribed at a national level.

Heating

C1 The main type of space-heating system is surveyed; more than one alternative is possible.

C2 The rated power is established for the main types of space heating. Information on age (or estimated age) should be available, primarily serving as a basis for boiler efficiency default values. Also, the meter or energy supplier account (subscription) should be included so as to identify which fuel is used where. For solar heating units, the total collector area should be assessed.

C3 If the systems have been tuned, information on primary efficiencies ought to be collected. If not, default values from national codes must be used, for example on the basis of type and age.

C4 Individual boilers may use different fuels. These must be assessed. A convenient way of finding out the preliminary distribution of the fuels is by asking the representative for estimated values. This may reveal values for fuels that are not billed.

C5 Certain boilers using different fuels have automatic change-over between types. Patterns may be revealed here.

C6 Supplementary heat sources are assessed. In some cases, these are not billed and the auditor has to find the energy consumption on the basis of estimates. Another problem is if, for example, an electrical cooker is used for heating (for example, if the heating system is shut off when nobody is home but both are turned on upon arrival). The cooker, which is normally a climate-independent energy consumer, is now a climate-dependent heater.

C7 Does one or more heat pumps exist? This heat pump may be a central unit or local units may be used for heating and cooling.

C8 The types of use of the heat pumps should be included.

C9 Whether the energy (electricity) supplied only to the heat pump is sub-metered. If so, the meter number should be filled in.

C10 A general question to obtain information on the main type of heat distribution.

C11 The last occasion on which the heating system was tuned. This is to provide information for C12.

C12 Information on the temperatures and flows in the heat distribution system, which can be assessed if appropriate information is provided under C11. If there are meters present, these can be read.

C13 The main types of heaters used (more than one alternative).

C14 Information on district heating.

C15 If tuning records are available, information should be gathered from these documents on the primary efficiencies of systems.

Cooling

Cooling often uses electricity, but can in a few cases use natural gas. District cooling is not common and is, for example, in Nordic countries only used for offices. However, it is included in the protocol as a future alternative.

D1 How the audited building is mainly cooled. Alternative 6 states no cooling. However, alternative 4 indicates the use of fans (in ceilings) without cooling. The use of fans is climate-dependent and their utility is considered to be greater during summer than during the winter season.

D2 Rated powers of the various types are noted, together with the age and the 'fuel'. The power of local units is summed. The age is an average value for unit ages.

D3 If a tuning protocol exists, the cooling efficiency factor is noted, if possible. Manufacturer information is also useful.

Domestic hot water

The use of domestic water varies considerably between households. It is seldom that energy for heating domestic water is sub-metered. For the purpose of determining this energy, as in the case where the same fuel and boiler is used for space heating, or with solar heating, it is advised that the partitioning of seasons is considered first. If this is not possible, water consumption data can be used to estimate the energy demand.

E1 Is there a central domestic hot-water boiler?

E2 Are meters for hot-water consumption available? Usually, they are not.

E3 The table is used to assess how hot-water heating is done for both the heating and cooling seasons. These may be different or there may be a combination of different types.

E4 The rated power and age of any separate water boiler.

E5   The rated power and age of the central water boiler (for multi-family building).

E6   The set-point temperature of the hot water is assessed, either by a reading of gauge or from what is stated in manuals.

E7   The size of the hot-water tank (measured or estimated).

Appliances

Most household appliances and lighting are electrical, and therefore the consumption is a part of an energy bill. With few exceptions, freezers and refrigerators are electrical unless gas is used. The most common appliance that might not be electrical is the cooker (hob and oven). For this reason, only two questions are formulated in this section.

F1   What types of fuel do the hob and oven use.

F2   How many hot meals are on an average cooked each day at home? From this frequency, a default value for the energy used for heating a meal is used.

Electricity meter information

Buildings may be equipped with more than one electrical meter. For this reason, it is important to assess which meter is measuring the various electrical uses.

G1   Assessment of the use of electricity in other buildings if this is registered by the electricity meter.

G2   Electricity meter number or code.

G3   Are any of the meters measuring specific energy use and which specific energy use is being measured by each meter?

G4   Are there external electrical devices that are not a part of the energy balance of the building?

G5   A list of external electrical devices. Energy consumption by these is estimated by finding the rated power of each device, multiplied by the run time when these are utilized. Other estimation types are possible.

Suppliers

The energy and water suppliers are to be listed. The information required is the names, addresses and phone numbers of the suppliers. The supplier account (or subscription) number of the customer (resident), the meter number, the fuel type and the unit of the quantity delivered must also be listed. The information on whether or not the billed quantity is estimated or exact for the billed periods must also be filled in. The conversion factor from the quantity unit to kWh should be noted.

In the third table, the delivery dates and quantities according to bills from the supplier (number according to the first table) are recorded.

REFERENCES

1. EN ISO 13790: 2004 Thermal Performance of Buildings – Calculation of Energy Use for Space Heating.

2. Kotsaki E and Sourys G, 2000, *Critical Review and State of the Art of the Existing Rating and Classification Techniques*. Report, Group Building Environmental Studies, University of Athens, Greece.

3. Vekemans G, Loncour X, Bradfer F and Crabbé C, 2000, *National Report of the SAVE Belas Project: Belgium – Final Report*, Study for the European Commission DG TREN Contract XVII/4.1031/Z/99-261. Vito Center 2001/ETE/001-07, Belgium.

4. Norlén U and Andersson K (eds), 1993, *The Indoor Climate in the Swedish Housing Stock*. Document D10:1993, Swedish Council for Building Research, Stockholm, Sweden.

5. ISO 9060:1990 Solar Energy – Specification and classification of instruments for measuring hemispherical solar and direct solar radiation. ISO/TC 180/SC1. International Organization for Standardization, Geneva, Switzerland.

6. Vallenor U and Wikström L, 1984, *Energiförbrukning i Byggnader – Delrapport 4: Mätning av oljeförbrukning och temperature*, Meddelande M84:17, The National Swedish Institute for Building Research, Gävle, Sweden (in Swedish).

7. Westergren K-E and Waller T, 1998, *Virtual Housing Laboratory – A system for simulating the energy use for heating in single family houses*, Working Paper No. 1, University of Gävle, Sweden.

8. Westergren K-E, Högberg H and Norlén U, 1999, 'Monitoring energy consumption in single-family houses', in *Energy and Buildings*, 29, 247–257.

9. Sálkever DS, 1976, 'The use of dummy variables to compute prediction errors, and confidence intervals, *Journal of Econometrics*, 4, pp 393–397.

10. Tolstoy N, Borgström M, Högberg H and Nilsson J, 1993, *Bostadsbeståndets tekniska egenskaper – ELIB-rapport nr 6*, Forskningsrapport TN:29, The National Swedish Institute for Building Research, Gävle, Sweden (in Swedish, available in English).

11. Energy Information Administration, 1997, *A Look at Residential Energy Consumption in 1997*, DOE/EIA-0632(97), Energy Information Administration, Office of Energy Markets and End Use, US Department of Energy, Washington DC. See also http://www.eia.doe.gov/emeu/recs/contents.html.

12. Tolstoy N, 1994, *The Condition of Buildings – Investigation methodology and applications*, TRITA-BYMA 1994:3, Doctoral dissertation. Department of Building Materials, KTH, Stockholm, Sweden.

13. Álvarez S and Gonzlez Falcon R, 2000, *Thoughts on Rating Methodology*, Escuela Superior de Ingenieros, DIE-Grupo Termotecnia, Universidad de Sevilla, Camino de los descubrimientos s/n, Sevilla, Spain.

14. EN 304: 1992. Heating boilers – Test code for heating boilers for atomizing oil burners.

15. Martin S, Wouters P and L'Heureux D, 1996, *Evaluation of a simplified method for the energy certification of non-occupied buildings*. Final Report, Save Help Project, Belgian Building Research Institute (BBRI), Brussels.

Further reading

Norlén U, 1982, *Temperaturundersökningen 1982 – Huvudinstruktion*, PM 1982-01-29. Statens institut för byggnadsforskning, Gävle, Sweden (in Swedish).

Wouters P and Loncour X, 2000, EURO-CLASS - Position paper of the BBRI about the objectives of the project. Division of Building Physics and Indoor Climate, Belgian Building Research Institute (BBRI).

CHAPTER 3
Energy normalization techniques

JAN AKANDER, SERVANDO ALVAREZ AND GUÐNI JÓHANNESSON

INTRODUCTION

Normalization of energy consumption can be done for several reasons and by different means. For the classification of buildings, the choice has been made to normalize energy use from three possible points of view. The objective of the three normalization types takes into consideration what input is available (specific energy use and temperatures) and for the reason that normalization is done.

Normalization takes into account:

- the size of the building
- the external environment climate
- the internal environment climate.

The size of the building is taken into consideration so as to create some variable that will allow comparisons of building energy performance. Within this task, the choice was to use the heated floor area as the benchmarking criterion. The use of heated floor area will relate energy use to the parts of the residence that are utilized and conditioned.

Within the framework of the methodology, it is advisable that the normalization with regard to building size is uniform for all countries. However, national customs already offer a diversity of normalization parameter types. The framework has, therefore, been left open for this parameter to be defined at a national level. The parameter, now the heated floor area, can be changed to the protected volume, a characteristic length etc. The use of national parameters, and thereby national energy use values, allows comparisons within a single country. Use of size conversion factors will allow comparison across national borders. For example, the conversion factor between the heated floor area and the heated volume is building height. The present parameter, heated floor area, illustrates how such a parameter is defined and utilized within the framework.

Normalization with respect to the external environment climate takes into consideration annual variations. By placing the building within a reference, or normal, climate for the site, not only will the performance of the building be comparable with the performance at other times, but it will also be possible to compare the performance with that of other buildings subject to the same external climate. For a building that has undergone only normalization with respect to the external environment it is assumed that the same indoor conditions prevail.

Normalization of the internal environment climate can be done in terms of a set of predefined indoor conditions. This type of normalization will, in fact, standardize a part of the end-user behaviour and will allow comparison between different occupants in the same residence.

In accordance with the definitions of Chapter 2, the energy concepts described below are used throughout the chapter.

Space-heating energy

Space-heating energy is the sum of the specific energy for space heating on an annual basis.

DELIVERED SPACE HEATING

The delivered space heating energy is the sum of heat dissipated from heating units $Q_{spaceheating}^{delivered}$.

SUPPLIED SPACE HEATING

Dividing delivered energy by the efficiency of the discharging unit gives the sum of supplied space-heating energy $Q_{spaceheating}^{supplied}$.

Space-cooling energy

Space-cooling energy is the sum of the specific energy for space cooling on an annual basis.

DELIVERED SPACE COOLING

The delivered space cooling energy is the sum of heat absorbed by cooling units $Q_{spacecooling}^{delivered}$.

SUPPLIED SPACE COOLING

Dividing delivered energy by the efficiency of the discharging unit gives the sum of supplied space-cooling energy $Q_{spacecooling}^{supplied}$.

Global energy

Annual global energy is the sum of specific energies. How external energy should be handled deserves special attention.

DELIVERED ENERGY

Delivered global energy is the sum of delivered specific energy and is straightforward in this sense. However, the delivered external energy $Q_{external}^{delivered}$ must be handled correctly. If this energy is delivered outside the considered building envelope, $Q_{external}^{delivered} = 0$ since by definition, this is not delivered into the building space and will not affect the heat balance of the building. In the case where $Q_{external}^{delivered}$ is delivered within the building envelope, but outside the considered apartment, it will be included in delivered global energy.

As an example, consider an apartment that is to be rated in a multi-family building. The apartment is billed for the energy use within the apartment, but it will also be billed an extra fee for shared costs for the heating of corridors and staircases and for elevator operation. This extra fee corresponds to $Q_{external}^{delivered}$, which is delivered within the envelope of the multi-family building and which will be included in the delivered global energy. However, if an outdoor pool is regularly heated, the external energy delivered to the pool will be considered to be zero $Q_{external}^{delivered} = 0$.

SUPPLIED ENERGY

Supplied energy is the sum of all supplied specific energies, and this also includes all energy supplied to external devices.

NORMALIZATION OF BEP AND MEP OUTPUT

The proposed methodology provides a normalization platform that depends on which experimental protocol has been employed. Primarily, it is the availability of data input that determines which types of normalization can be applied and the principles of the rating scheme. Results from BEP and MEP are treated in the same way, so as to make the normalized results from different residences comparable. Both are normalized with respect to the heated floor area and the external climate.

The reason for excluding normalization of the internal climate in the rating procedure is twofold:

- BEP does not always offer reliable information on internal environment conditions.
- The rating scheme is based on normalized energy use with respect to the building size (floor area) and the external climate. The indoor conditions and consequently energy use are dependent on occupancy behaviour. Low-energy consumers, who prefer to have a low indoor temperature, should not be 'punished' in the rating process. In contrast, high-energy consumers should receive a worse rank. The rating method evaluates energy use of the building including the effect of occupancy behaviour.

Because the indoor temperature is assessed in MEP, normalization of internal climate can be performed together with the other normalization types. The outcome of this normalization will not be rated and should thus inform the occupants how their behaviour compares with that of 'standard occupants'.

BEP

The Billed Energy Protocol (BEP) makes use of bill information to serve as input to the rating procedure.

Complementary information that may be obtained is as follows:

- the heated floor area assessed from the audit and the plans (or other necessary data to determine the size parameter)
- indoor temperatures obtained from the occupants or from representatives of the maintenance operator
- set-point temperature of the main energy conversion systems and energy distribution devices, if available
- external environment climate for the period covered by bills; this may be obtained from the closest climate station
- reference external environment climate, which is determined at a national level.

Normalization of BEP output values is done on the basis of the heated floor area and the external climate only. Normalization is not done for the internal environment because the indoor temperatures are considered to be unknown. End-user information may be erratic, and the situation may be such that set-point temperatures of the heating and cooling units are not accessible.

MEP

The Monitored Energy Protocol (MEP) offers an insight into the internal environment temperature during the season that measurements were conducted, although the monitoring period may be considered short from the viewpoint of the energy involved. As well as information from the measurements and the audit, information that can be assessed elsewhere is as follows:

- the heated floor area assessed from the audit and the plans (or other necessary data to determine the size parameter)
- indoor temperatures obtained from the occupants or from representatives of the maintenance operator and from the monitoring period
- set-point temperature of the main energy conversion system and the energy distribution devices, if available
- external environment climate for the period covered by bills; this may be obtained from the closest climate station – monitored external environment variables can be compared with those from climate stations, so that systematic deviations can be detected
- reference external environment climate, which is determined at a national level.

Normalization of MEP output values is done on the basis of the heated floor area and the external climate, in the same manner as for BEP outputs. This makes normalized results from the two protocols comparable. Moreover, with the insight into internal climate temperature, normalization of internal climate can be performed as a service to the occupants. This gives the occupants information on their energy use compared to that of 'standard occupants' or to 'standard indoor conditions'.

NORMALIZATION WITH RESPECT TO THE EXTERNAL ENVIRONMENT CLIMATE

Normalization with respect to the external environment climate is performed for both BEP and MEP. This normalization is preparatory to the rating procedure: the residence is adjusted for the reference climate, which will bring changes to space-heating and space-cooling energy. The other specific energies may also receive normal values, for example default values for a 'standard' or 'normal' energy use for a normal occupancy. This type of normalization cannot be done for the output from BEP, the primary reason being that the temperatures of the internal environment are not known. Normalized values of this type influence the internal environment and can only be done within MEP, when monitoring of temperatures has been performed. Therefore, the internal climate is not normalized prior to the rating procedure. The internal climate conditions during the prevailing year are assumed to be the same as those for the reference year. Implicitly, climate-independent variables are assumed to be the same for the two time periods under consideration.

Three normalization procedures are proposed, and the choice of normalization method (or methods) to be used will be determined at a national level. These three are:

- the heating and cooling degree-day method (DDM)
- the modified utilization factor method (MUFM)
- the Climate Severity Index (CSI).

The heating and cooling degree-day method

The heating and cooling degree-day method is a traditional method that has been in use for decades, in both the academic and the industrial worlds. It is a well-known method and is easy to use, although there are several variants.

In general, the concept primarily builds on the temperature difference between a base indoor temperature and the outdoor temperature, multiplied by the duration of the temperature difference. It is quite common for the length of heating and cooling season to be pre-determined. The base indoor temperature is also prescribed, with different values and definitions in various countries.

Expressed in equation form, the number of degree-days calculated on a daily basis, are:

- for the heating season

$$HDD = \sum_{t_{start}}^{t_{end}} (\theta_{HDD\_base} - \theta_e) \tag{3.1}$$

- for the cooling season

$$CDD = \sum_{t_{start}}^{t_{end}} (\theta_e - \theta_{CDD\_base}) \tag{3.2}$$

Here, the symbols denote:

HDD = degree-days for heating [°C·days]
CDD = degree-days for cooling [°C·days]
$\theta_{HDD\_base}, \theta_{CDD\_base}$ = base indoor temperatures for the considered period [°C]

θ_e = mean outdoor temperature on a daily basis [°C]
t_{start} = start day of the season
t_{end} = end day of the season.

The following remarks should be noted:

- The numeric values of θ_{HDD_base}, θ_{CDD_base} and the definition of the length of the heating/cooling season are left to be determined at a national level.
- The degree-days are calculated on a daily basis.
- The summing is done for non-negative differences, which indicate the presence of a heating or cooling requirement.
- Such calculations are made for the year with the actual climate (HDD and CDD), and for the year with the reference conditions (HDD^N and CDD^N).

The most straightforward method of normalization based on the degree-day concept is to use the ratio between the reference and the prevailing degree-days. The following expressions are used:

$$Q_{spaceheating}^{N\,delivered} = Q_{spaceheating}^{delivered} \cdot \frac{HDD^N}{HDD} \tag{3.3}$$

$$Q_{spacecooling}^{N\,delivered} = Q_{spacecooling}^{delivered} \cdot \frac{CDD^N}{CDD} \tag{3.4}$$

LIMITATIONS OF THE MODEL

The degree-day concept has several limitations:

- First of all, only temperature is taken into consideration. For instance, there is no solar radiation taken into account.
- There is also a problem with defining the magnitude of the base indoor temperature and the length of each season for individual residences.
- The use of degree-days is considered to be adequate if internal gains and solar gains do not significantly influence the heat balance of the building. This is only the case for Nordic countries, when the space heating energy forms a large part of the whole energy use.

The cooling degree-day method has severe limitations in that solar radiation is not accounted for and that cooling is a non-linear phenomenon. The degree-day concept assumes that the heat-loss factor is constant and that heat loss is proportional to the temperature difference. In most cases, cooling is only applied in a few rooms and for a limited period of time, in general during the peak hours during the day.

TOLERANCES OF THE MODEL

The normalized values that are obtained should be tagged with an error band. For the degree-days, three calculations must be performed. These are, for the example of heating, degree-day calculations for the base temperature θ_{HDD_base} and for $\theta_{HDD\_base} \pm 3°C$. This applies to both the prevailing and the reference climates. The results of normalization can be given within the bandwidth of:

$$\min Q_{spaceheating}^{N\,delivered} < Q_{spaceheating}^{N\,delivered}(\theta_{HDD\_base}) < \max Q_{spaceheating}^{N\,delivered}$$
(3.5)

where

$$\min Q_{spaceheating}^{N\,delivered} = \min[Q_{spaceheating}^{N\,delivered}(\theta_{HDD\_base} + 3),$$

$$Q_{spaceheating}^{N\,delivered}(\theta_{HDD\_base} - 3)]$$

$$\max Q_{spaceheating}^{N\,delivered} = \max[Q_{spaceheating}^{N\,delivered}(\theta_{HDD\_base} + 3),$$

$$Q_{spaceheating}^{N\,delivered}(\theta_{HDD\_base} - 3)]$$

Analogously, the same types of equations apply to space cooling.

The modified utilization factor (MUF) method

The modified heating degree-day method was proposed for two reasons:

- The normalization method has to take into consideration the solar gains.
- The set-point temperature of the main heating system has to be used since this is a temperature that can be assessed from an audit, without the need of detailed measurements.

The space-heating requirement, as specified earlier, is dependent on four components, heat delivered into the space by heating systems, appliances and lighting, solar and metabolic heat. On the basis of a steady-state approach, the heat losses by means of transmission and ventilation will equal the heat delivered into the space, such that

$$Q_{heat\,losses} = Q_{spaceheating}^{delivered} + Q_{appliances}^{delivered} + Q_{metabolic} + Q_{solarheat}$$
(3.6)

The indoor temperature that is obtained over time is a result of when, where and how much energy is delivered into the space, the heat-loss factor and the heat capacity of the residence. The indoor temperatures of thermal zones in the building are usually unknown during the season under consideration. The most reliable temperature data that can be assessed during the heating season, without monitoring, is the set-point temperature of the heating system. Therefore, calculations that serve to determine delivered space-heating energy, both actual and normalized, will make use of the set-point temperatures. A utilization factor for internal and solar heat gains is introduced. By means of the utilization factor and adjusting the indoor temperature to be the set-point temperature, normalization of the space-heating energy is possible.

INTRODUCTION OF THE UTILIZATION FACTOR FOR HEAT GAINS

If the indoor temperature is not known, the set-point temperature of the heating system can be used in the calculations. The calculation has its roots in prEN ISO 13790,[1] which makes use of the so-called utilization factor for internal heat gains. The utilization factor is a measure of

the part of the internal heat gains that is useful in obtaining the set-point temperature within the space. The rest of the heat gains are considered to create an internal temperature that is above the set-point temperature.

A re-formulation, in which the space temperature is now the set-point temperature (*spt*) rather than the actual temperature of the space, can be made such that

$$Q_{heatlosses}^{at\_spt} = Q_{spaceheating}^{delivered} + \eta_{UF}(Q_{appliances}^{delivered} + Q_{metabolic} + Q_{solarheat})$$
(3.7)

The utilization factor η_{UF} is determined either on a seasonal or a monthly basis, depending on the periodicity of bills that have been collected.

Note the differences in energy losses, $Q_{heatlosses}^{at\_spt}$ and $Q_{heatlosses}$. These are related to the temperatures that are used in the calculations, where

$$Q_{heat\,losses}^{at\_spt} = H(\theta_{spt} - \theta_e) \cdot t$$
(3.8)

$$Q_{heat\,losses} = H(\theta_i - \theta_e) \cdot t$$
(3.9)

with H denoting the heat loss factor of the building, t_s the period of time, θ_e the mean external temperature, θ_{spt} the set-point temperature of the heating system and θ_i the mean indoor temperature.

The procedure of prEN ISO 13790 calculates heat losses relative to the set-point temperature. The procedure makes use of the so-called gain–loss ratio γ during a considered period of time, defined such that

$$\gamma = \frac{Q_{appliances}^{delivered} + Q_{metabolic} + Q_{solarheat}}{Q_{heat\,losses}^{at\_stp}}$$
(3.10)

The utilization factor is calculated using the gain–loss ratio so that

$$\eta_{UF} = \frac{1 - \gamma^a}{1 - \gamma^{a+1}} \quad \text{if} \quad \gamma \neq 1$$
(3.11)

$$\eta_{UF} = \frac{a}{a + 1} \quad \text{if} \quad \gamma = 1$$
(3.12)

where

$$Q_{heatloss}^{required} = \sum Q_{spaceheating} + \eta_{UF}\left(\sum Q_{appliances} + \sum Q_{solarheat}\right.$$
$$\left. + \sum Q_{metabolic}\right) \quad \text{for month-wise calculations}$$

$$a = 0.8 + \frac{\tau}{28} \quad \text{for seasonal calculations}$$

The time constant τ is defined as the ratio of the effective heat capacity to the heat-loss factor of the building.

NORMALIZATION OF SPACE HEATING ENERGY

Normalization of the space heating energy can be performed with the utilization factor concept as presented above. Some modifications have to be made, since a 'backward' prEN ISO13790 procedure is applied. The steps of this procedure are illustrated below.

prEN ISO13790 is based on a descriptive model, which assumes that characteristics and data on the building, the energy systems, end-user behaviour and occupancy are known or prescribed, together with the reference external climate. From the viewpoint of this methodology framework,

Table 3.1 Time constants (hours) for various building types with consideration of the internal mass. The table may be extended by evaluating the amount of insulation. The values in this table are only for illustrative purposes

Building type	Light-weight	Medium-weight	Heavy-weight
Single-family house	20	30	60
Terrace house	25	40	80
Apartment	30	50	90

it is the output of this descriptive model that has been assessed in terms of billed quantities, for the actual climate.

Background information from the audit

The auditor chooses the time constant of the residence at the time of audit. Table 3.1 shows the alternatives, and the magnitudes of the time constants are to be taken from representative buildings in a particular country.

From the audit, information is gathered for the solar apertures and the orientations. Based on national or regional areas, there must be computational codes for performing solar radiation calculations. It may be convenient to use a development of national standard codes for this purpose, or to adapt routines from building simulation programs.

Estimation of heat losses at the prevailing temperature

At the prevailing indoor temperature, an estimation of the heat losses by transmission and ventilation is made. The estimated right-hand side variables for a month-wise or seasonal period will give

$$Q_{heat\ losses} = Q_{spaceheating}^{delivered} + Q_{appliances}^{delivered} + Q_{metabolic} + Q_{solarheat} \tag{3.13}$$

where

$Q_{spaceheating}^{delivered}$ is the estimated sum of the delivered space heating energy from conversion systems (from BEP and MEP)

$Q_{appliances}^{delivered}$ is the estimated sum of the energy delivered from appliances and conversion systems (from BEP and MEP)

$Q_{metabolic}$ is the estimated energy delivered from the occupants calculated from the dissipated heat per occupant and the time of presence

$Q_{solarheat}$ is the estimated solar energy delivered into the building, based on climate data calculations (national methods) and solar aperture estimations (from BEP and MEP).

Calculation of the modified gain–loss ratio

The modified gain–loss ratio is now calculated for the period considered. The calculation is iterative, assuming an initial value for $\eta_{UF}^{i=1}$. This value can be set to equal unity and five iterations are enough. The iteration will loop through

the following equations for iteration step i:

$$\gamma^{i+1} = \cfrac{1}{\cfrac{Q_{heat\ losses}}{Q_{appliances}^{delivered} + Q_{solarheat} + Q_{metabolic}} - (1 - \eta_{UF}^i)} \tag{3.14}$$

Then,

$$\eta_{UF}^{i+1} = \frac{1 - (\gamma^{i+1})^a}{1 - (\gamma^{i+1})^{a+1}} \quad if\ (\gamma^{i+1}) \neq 1 \tag{3.15}$$

$$\eta_{UF}^{i+1} = \frac{a}{a+1} \quad if\ (\gamma^{i+1}) = 1 \tag{3.16}$$

The computational loop will give the final value for η_{UF}^*.

Determining the potential heat losses at the set-point temperature

The final value of η_{UF}^* will now be used to estimate what the heat losses would be if the internal temperature were equal to the set-point temperature. Expressed in equation form, this corresponds to

$$Q_{heat\ losses}^{at\_spt} = Q_{spaceheating}^{delivered} + \eta_{UF}^*(Q_{appliances}^{delivered} + Q_{metabolic} + Q_{solarheat}) \tag{3.17}$$

Heat losses for the reference year

The next step is to normalize heat losses with respect to the external climate. The base indoor temperature is in this case the set-point temperature. This is more or less a traditional degree-day calculation for month-wise or seasonal average temperatures, as follows

$$Q_{heat\ losses}^{N\ at\_spt} = Q_{heat\ losses}^{at\_spt} \cdot \frac{(\theta_{spt} - \theta_e^N)}{(\theta_{spt} - \theta_e)} \tag{3.18}$$

Normalization of space-heating energy

From this step on, the calculations are the traditional prEN ISO 13790 calculations. However, preparatory calculations are performed to estimate the solar energy for the reference climate that is delivered into the building (and, if required, normalized values from appliances and metabolic heat). The gain–loss ratio for the normalized conditions is now

$$\gamma^N = \frac{Q_{appliances}^{delivered} + Q_{metabolic} + Q_{solarheat}^N}{Q_{heat\ losses}^{N\ at\_stp}} \tag{3.19}$$

The normalized utilization factor is calculated using the normalized gain–loss ratio, here expressed as

$$\eta_{UF}^N = \frac{1 - (\gamma^N)^a}{1 - (\gamma^N)^{a+1}} \quad if\ (\gamma^N) \neq 1 \tag{3.20}$$

$$\eta_{UF}^N = \frac{a}{a+1} \quad if\ (\gamma^N) = 1 \tag{3.21}$$

The normalized space-heating energy is calculated from the set of variables as follows

$$Q_{spaceheating}^{N\ delivered} = Q_{heat\ losses}^{N\ at\_spt} - \eta_{UF}^N(Q_{appliances}^{delivered} + Q_{metabolic}^N + Q_{solarheat}^N) \tag{3.22}$$

This is the normalized delivered space-heating energy. The supplied space-heating energy for each fuel type is the part that it has contributed to the specific energy for space heating so that

$$Q^{N\,supplied}_{spaceheating} = \frac{Q^{N\,delivered}_{spaceheating}}{\eta_{sh}} \qquad (3.23)$$

NORMALIZATION OF SPACE-COOLING ENERGY

As in the case of heating, the cooling season involves an energy use for maintaining a desired internal climate. However, even in the most idealized cases the quantity that is used is non-linear, which makes assessment and normalization more complicated. Though the heat balance equation (see below) initially seems straightforward, determination of each variable becomes exceedingly difficult to master with low inaccuracy:

$$Q^{delivered}_{spacecooling} + Q_{heat\,losses} = Q^{delivered}_{appliances} + Q_{metabolic} + Q_{solar} \quad (3.24)$$

Furthermore, space cooling is seldom applied to the whole residence. Use is made of cooling in rooms that are occupied, and usually only for a limited time during the occupants' presence. For these reasons, space cooling will strictly be limited to analysis of bills (supplied energy) and a different normalization scheme will be applied to that used in the case of space heating.

At the present time, there is no European norm that covers the space cooling energy requirement in the same way as prEN ISO 13790 does for space heating. Such procedures are under development and similarly make use of the utilization factor, for example the Dutch building code NEN 2916 (for commercial buildings).[2]

LIMITATIONS OF THE MODEL

The modified utilization factor method has several limitations. First, a central set-point temperature may not be available in the building considered. If so, national default values will have to be applied, preferably taking into consideration information from the occupants or by measuring temperature profiles during the day.

Another limitation is the estimation of solar radiation that is delivered through apertures. Shading factors, curtain factors and user behaviour have great impact on solar irradiation. Not only do these vary over the seasons, but they also vary considerably throughout each day. However, within the frame of normalization calculations, the effect of an erroneous solar term will only affect the normalization of the heat loss; the same error will afterwards be subtracted from the final results (see Equations 3.17 and 3.22).

Finally, the use of a utilization factor can be doubtful. In this application, the time constant is unknown and will be estimated on the basis of the auditor's intuition (calculations based on plans would be quite cumbersome). Moreover, the underlying theory of the utilization factor is based on generalization of data from numerous buildings, which makes results for one specific building uncertain. On the other hand, based on the limitations within the experimental protocols BEP and MEP, the order of the deviations that may arise from this technique will probably not be greater than the uncertainties within the specific energies.

TOLERANCES OF THE MODEL

The tolerances of the model are not trivial to assess in this method. This is because many of the variables are linked to each other, as will be shown in the following text. The calculations for tolerances will have to be specified on a national basis; since the influence of climate variables in this model will be non-linear. Below, an example is given, but the parameters will have to be changed to relevant national values.

If the set-point temperature is assessed during the audit, the calculations of 'degree-months' can be performed such that the base temperature corresponds to θ_{spt} and for $\theta_{spt} \pm 1°C$. If the set-point temperature is not known, the 'degree-month' part of the calculations should be done for the base temperature $\pm 3°C$.

A sensitivity analysis must be done on the basis of the time constant, with $\tau \pm 25\%$.

The influence of solar radiation must be assessed where the solar apertures vary in size, $A_{s,n} \pm 50\%$ or where $Q_{solarheat} \pm 50\%$, in combination with the base temperature deviations.

The results of normalization can be given within the bandwidth of

$$\min Q^{N\,delivered}_{spaceheating} < Q^{N\,delivered}_{spaceheating} < \max Q^{N\,delivered}_{spaceheating} \qquad (3.25)$$

where

$$\min Q^{N\,delivered}_{spaceheating} = \min[Q^{N\,delivered}_{spaceheating}(\theta_{spt} \pm 3,$$
$$Q_{solarheat} \pm 50\%, \tau \pm 25\%)]$$

$$\max Q^{N\,delivered}_{spaceheating} = \max[Q^{N\,delivered}_{spaceheating}(\theta_{spt} \pm 3,$$
$$Q_{solarheat} \pm 50\%, \tau \pm 25\%)]$$

Analogously, the same types of equations would be used in space-cooling applications.

A pre-study

The influence of each variable is not trivial, since these influence one another. A pre-study is given in this section, showing how the variables affect the outcome of the procedure, which is the space-heating energy. In order to facilitate the presentation, some definitions are given below.

The relative change in space heating is based on the space-heating energy that is assessed, whereas the normalized space-heating energy is the outcome of the MUF method. The relative change that the normalized space heating gives rise to in comparison to the actual space-heating energy is formulated such that

$$change = \frac{Q^{N\,delivered}_{space\,heating} - Q^{delivered}_{space\,heating}}{Q^{delivered}_{space\,heating}} \qquad (3.26)$$

After some mathematical operations, the following expression can be obtained:

$$change = \frac{\beta - \eta^{N}_{UF} \cdot \rho \cdot \gamma}{1 - \eta_{UF} \cdot \gamma} \qquad (3.27)$$

where

β = the ratio between temperature differences $\frac{\theta_{spt} - \theta_e^N}{\theta_{spt} - \theta_e}$

ρ = the ratio between heat gains, $\frac{Q_{gains}^N}{Q_{gains}}$

γ = the gain–loss ratio of the actual period, $\frac{Q_{gains}}{Q_{losses}^{at\_spt}}$

and $\eta_{UF}(\gamma, \tau)$ and $\eta_{UF}^N(\gamma^N, \tau)$ are dependent on the time constant τ and the gain–loss factors of the actual and the normal periods. The relationship for the gain–loss factors is that $\gamma^N = \frac{\rho}{\beta}\gamma$.

The normalized space heating will be roughly dependent on the gain–loss ratio for the actual period, the ratio of heat gains, the temperature difference ratio for the two periods considered and the time constant. In a computational study, these sets of equations were used to derive the order of magnitude of the change in space-heating energy.

The first set of calculations was intended to show the influence of the temperature of the external environment. The ratio between heat gains ρ was set to unity. The results are shown in Figures 3.1, 3.2 and 3.3, where the time constant has been given the value of 30, 50 and 70 hours respectively. Calculations were based on the heating requirement for the heating season (not on a month-wise basis).

Figure 3.1 shows the relative change in space-heating energy as a function of the gain–loss ratio for the actual year and the variation of the temperature difference ratio. For the case where the normal year is the same as the actual year, β is unity and the relative change is zero for all γ. (By dividing γ by β, β can be substituted by γ^N on the x axis.) For a colder normal year, β increases and will lead to a larger space-heating requirement than for the actual year, since this means a reduction in γ^N. If the γ value is followed with changing β, the corresponding change in space heating can be observed.

A case in Sweden is taken as an example; the gain–loss ratio for the heating season is of the magnitude of 0.7. This involves some ±70% if the temperature difference ratio varies ±50%. The temperature difference does not vary this much from one heating season to the other.

In Figure 3.2, the results for a building with the time constant 50 hours are presented. The relative influence of changes in the external temperature is larger than for the 'thermally lighter building'. This is because the time constant is a function of the thermal mass of the building and also of the level of insulation. A large time constant means more insulation and thus less absolute energy use when β is unity. The relative change in space heating will therefore increase for larger time constants, although the absolute change is less than for the cases with a smaller time constant.

The largest relative changes are found for large time constants and when the temperature of the outdoor environment has dropped and heat gains are large during the heating seasons. This effect is clearly seen in Figure 3.3, where the building time constant is 70 hours.

The same type of analysis has been done for the case where the ratio of heat gains ρ is varied and the temperature difference ratio is kept at unity (Figures 3.4, 3.5 and 3.6). In practice, the heat gains are, with the exception of solar

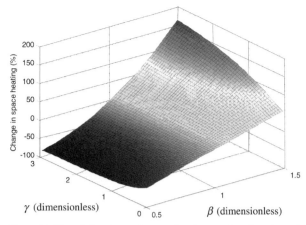

Figure 3.1 The relative change in space heating as a function of γ and β when ρ is kept at unity. The time constant of the building is 30 hours

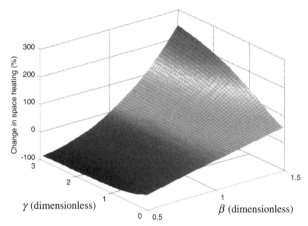

Figure 3.2 The relative change in space heating as a function of γ and β when ρ is kept at unity. The time constant of the building is 50 hours

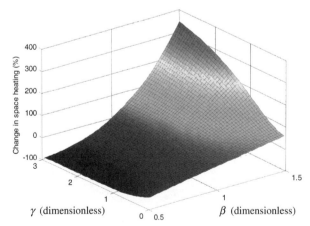

Figure 3.3 The relative change in space heating as a function of γ and β when ρ is kept at unity. The time constant of the building is 70 hours

gains, fairly constant. The ratio will reflect changes in solar gains, together with any climate-independent internal gains.

Again, the greatest relative changes in space heating are obtained with large values of γ in buildings with a large time constant. The largest impact of change in heat gains is found when the heat gains are decreased.

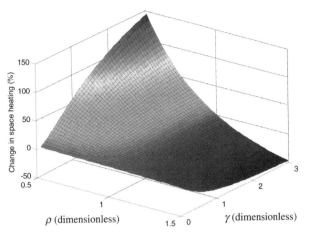

Figure 3.4 The relative change in space heating as a function of γ and ρ when β is kept at unity. The time constant of the building is 30 hours

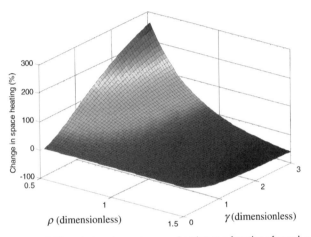

Figure 3.5 The relative change in space heating as a function of γ and ρ when β is kept at unity. The time constant of the building is 50 hours

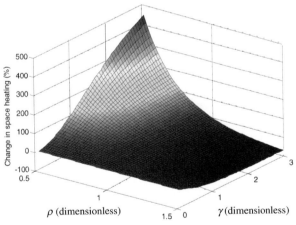

Figure 3.6 The relative change in space heating as a function of γ and ρ when β is kept at unity. The time constant of the building is 70 hours

Figure 3.6 shows that the gain ratio has the largest impact for buildings with a large time constant. A severe reduction in gain ratio leads to a large increase in space heating compared to the case of small time constants. This is because the building with a large time constant uses heat gains

more effectively, and when heat gains are reduced and limited, this will have to be compensated by space-heating energy.

The Climate Severity Index

In terms of the heating requirements of buildings, two climatic conditions can be considered to be 'identical' when the heating energy consumption of a certain building is the same under the two climatic conditions.

In this context, the concept climatic conditions must be understood in a broad sense:

- two actual years at the same location
- two actual years at different locations
- an actual year and the reference year at a certain location
- two reference years at two locations respectively
- and so on.

These definitions can obviously be translated to the cooling requirements and logically it can happen that two climatic conditions can be 'equal' in terms of heating requirements but 'different' in terms of cooling requirements. The reciprocal situation (equal for cooling and different for heating) is also true.

The idea can be extended by saying that, for a certain building, a climatic condition is x times more severe than another when the energy consumption of this building is x times greater under the former condition than under the latter.

One of the ways to characterize the climatic dependency of the heating or cooling requirements of buildings in the framework of the Euroclass project,[3] is the so-called Climatic Severity Index (CSI), which makes it possible to compare the 'severity' of different climatic conditions. The more severe the climate, the bigger the energy requirements of the buildings and, consequently, the bigger the value of the CSI.

There are, in principle, two CSIs, one for the heating season and other for the cooling season. The CSI as it is used is conceptually a positive number. When the CSI is zero or negative, it can be assumed that there are no significant energy requirements (for heating or cooling) under these climatic conditions.

DETERMINATION OF THE CSI

Let us suppose that the heating requirements are calculated for a given building under different climatic conditions that correspond either to different locations or to different years at a certain location or to a combination of both. The set of values obtained is divided by the value corresponding to a, say, pivot or representative climatic condition (see Table 3.2). The reduced heating requirements for this climatic condition will obviously become unity (1) and the reduced values for the other climatic conditions will provide the ratio between these and the heating requirements that would be obtained for the reference climate.

Table 3.2 Heating energy demands used in the calculation of the winter CSI

Burgos	**−21.373**	**1.9**
Soria	−18.427	1.6
Segovia	−15.945	1.4
Madrid	**−11.412**	**1.0**
Bilbao	−11.097	1.0
Barcelona	−7.782	0.7
Seville	−3.769	0.3
Cádiz	−2.362	0.2

In this way, *an index of the relative influence of the climate on the heating consumption of a building* is obtained. For instance, a climatic condition that yields a value of 0.6 implies that the heating requirements of the building are 60% of those of the pivot. A value of 1.4 means that the heating requirements are 140% bigger than in the pivot climatic condition.

The absolute climatic influence on the heating requirements of a building depends strongly on the building characteristics, but the relative climatic influence (as it has been defined above) is quite independent of factors such as the quality of the envelope, the window to wall ratio or the orientation of the building. The only significant factor remaining is the use of the building, mainly because of the internal gains.

With the assumption that buildings in the same sector (for instance residential) have a similar value for their internal gains, it is possible to calculate the CSI corresponding to each sector for a particular country or a wider geographic area.

The preparatory steps are:

1. selection of the pivot climatic condition
2. selection of a sample of the building typologies common in the geographic area
3. selection of climatic conditions covering the different climates of the area – typically typical meteorological year (TMY) or test reference year (TRY) of selected locations.

Next, the energy requirements of the buildings under different scenarios, for example the orientation of the main façade, different UA values, etc., are calculated via compu-

tational simulations. The results for every combination run for the selected climatic conditions are calculated and provide the CSI for each combination. The average of the CSIs for all the combinations is the CSI for this sector.

This resulting CSI is finally correlated with the conventional climatic variables corresponding to all the climatic conditions selected (degree-days, monthly average global solar radiation, insolation etc.) in order to make the concept applicable to climatic conditions different from those used to perform the simulations.

EXAMPLE AND ROBUSTNESS OF THE CSI

Two CSIs (one for heating and one for cooling) were calculated in 1996 during the project for developing the Spanish Energy Labelling for Social Housing (CEV project hereinafter).[4] These two indexes were obtained from simulations based on the 'Passport +' computer code. The CSI was applied to the capitals of the 50 Spanish provinces and the climatic condition used to make the values equivalent was the TRY of Madrid.

Figure 3.7 shows the correlation of the CSIs for summer and winter as a function of degree-days and the monthly average solar radiation.

The winter CSI correlation was obtained as follows:

$$CSI = a^*DD + b^*n/N + c^*DD^2 + d^*n/N^2 + e \qquad (3.28)$$

where

$DD =$ the degree-days for heating with a base temperature of 20°C for the months of January, February and December

$n/N =$ the ratio between the actual insolation hours and the maximum insolation hours for that latitude for the months of January, February, and December

$a = 2.395\text{E} - 03$
$b = -1.111\text{E} + 00$
$c = 1.885\text{E} - 06$
$d = 7.026\text{E} - 01$
$e = 5.709\text{E} - 02$
$R^2 = 0.99$

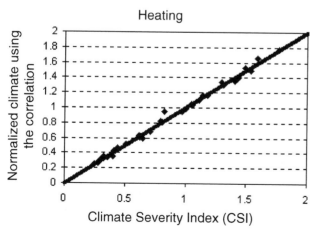

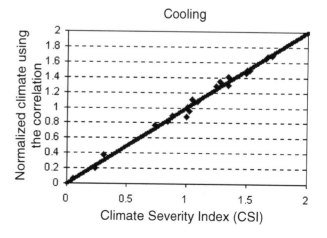

Figure 3.7 CSI correlation for winter and summer seasons

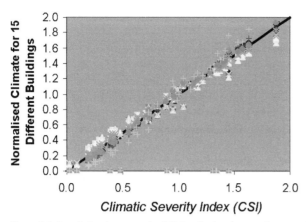

Figure 3.8 Correlation between the CSI for the heating requirement of 15 representative Spanish residential buildings in 50 locations

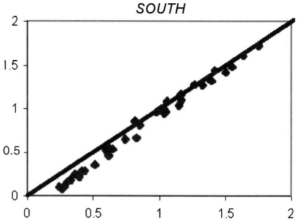

Figure 3.9 Correlation for a Belgian building in Spanish locations

The summer CSI correlation was obtained as follows:

$$CSI = a^*DD + b^*n/N + c^*DD^2 + d^*n/N^2 + e \qquad (3.29)$$

where

DD = the degree-days for heating with a base temperature of 20°C for the months of June, July, August, and September

n/N = the ratio between the actual insolation hours and the maximum insolation hours for that latitude for the months of June, July, August, and September

$a = 1.090E - 02$
$b = 1.023E + 00$
$c = -1.638E - 05$
$d = -5.977E - 01$
$e = -3.370E - 01$
$R^2 = 0.97$

In the Euroclass project, the applicability of these expressions was checked for three effects:

- the influence of the computer code used
- the influence of the building type
- the influence of the range of climatic conditions.

In all the cases, the CSI correlations developed for the CEV project proved to be good enough to characterize the influence of the climatic conditions on the heating or cooling requirements.

As an example, Figure 3.8 compares the CSI for heating the 50 Spanish locations (according to the CEV project) and the CSI obtained for 15 residential buildings representative of the Spanish typology and construction, using DOE-2E as the code for simulation. The quality of the fit demonstrates that the CSI is independent of the way of assessing the performance of the building. Consequently, it can be applied to any simulation-based or monitoring-based scheme of building energy assessment.

Figure 3.9 shows the effect of trying to apply the correlation to a building with a very different typology and construction standard (the Belgian house referred to as building 1 in reference [3]) when placed in the Spanish locations.

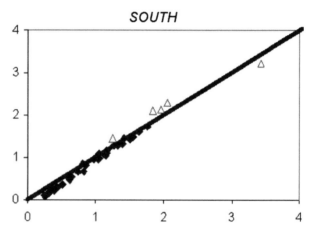

Figure 3.10 Correlation for a Belgian building in Spanish locations (filled symbols) and at the sites of Marseille, Paris, Lyon, Uccle and Stockholm (open triangles)

Figure 3.10 shows the effect of adding new climates (Marseille, Paris, Lyon, Uccle and Stockholm) represented with open triangles. The results are acceptable, except when the correlation is applied in Stockholm with a climate that is clearly colder than that of Spain.

In summary, the correlation developed in the CEV project is consistent when applied in a different climatic and constructive context and could be used to cover the majority of EU countries. However, it is obvious that a better approach would be obtained for CSI if, at a national level, a correlation is obtained based on each specific range of climates, building typologies and construction practices.

USE OF THE CLIMATE SEVERITY INDEX

The aim of the Climate Severity Index (CSI) in the Euroclass project was to predict the climate-dependent energy requirements of a building in a standard year from the results of energy requirements obtained in an actual year.

Multiplying the heating or cooling requirements (delivered energy) by the ratio between the reference CSI (superscript N) and the actual CSI produces this normalization. The following expressions (analogous to

Equations 3.3 and 3.4) are used:

$$Q_{spaceheating}^{N\,delivered} = Q_{spaceheating}^{delivered} \cdot \frac{CSI_{heating}^{N}}{CSI_{heating}} \tag{3.30}$$

$$Q_{spacecooling}^{N\,delivered} = Q_{spacecooling}^{delivered} \cdot \frac{CSI_{cooling}^{N}}{CSI_{cooling}} \tag{3.31}$$

NORMALIZATION WITH RESPECT TO THE INTERNAL ENVIRONMENT CLIMATE

Normalization with respect to the internal environment climate requires a set of conditions that specify what the internal climate is and, for the occupants, their rate of presence and their habits. In other words, the standard occupant has to be defined. The variables and parameters that will be influenced by standardization, described in a generalized manner, are:

- the temperatures and temperature control of the indoor climate
- energy use for heating standard tap-water consumption
- energy use for a standard set of household appliances and lighting
- a standardized method of shading control.

This task, deciding the choice of the set of variables and parameters, including the procedures for determining numerical values, should be determined by national interests. Application of normalization of the internal environment is only done within MEP, and the motivation for this was stated in the section 'Normalization of BEP and MEP output' earlier in this chapter.

The procedure for normalization of indoor conditions should be considered to be an extra service within MEP. The quantities of energy determined by this procedure will provide to the end-customer (the occupant(s)) information on how good their thermal behaviour is in comparison to that of the standard occupant, had the standard occupant lived in the residence considered. For example, how much more or less energy for tap-water heating is the occupant using in comparison to a standardized quantity?

Within this framework, only normalization of the temperature of the indoor environment will be considered. As stated above, normalizing values for other variables is left open for national initiatives to complete within the framework. The primary motivation for normalizing internal temperatures only is that these directly influence space-heating and cooling energy.

To some extent, normalization of internal environment is performed in the procedures of normalization with respect to external climate. This is implicit in the choice of internal base temperature in the degree-day method (DDM) and in relating the heat loss to the set-point temperature in the modified utilization factor method (MUFM). The Climate Severity Index is based on numerous simulations where the indoor conditions have been prescribed.

The methodology within MEP requires that temperatures in the thermal zones be measured. Measurements are made for periods of:

- some two weeks (short period)
- an extended period much longer than two weeks (long period).

The length of the monitoring period depends on the purpose of the measurement. The short period is to monitor the size of the specific energies and to record temperatures. The extended (long) period is primarily for assessing the heat-loss factor of the building by measuring delivered heat and temperatures.

The monitoring period and the choice of normalization method will influence how to determine energy consumption with normal internal and external conditions.

The heating and cooling degree-day method

The short-term monitoring campaign verifies the quantities of specific energy and the internal temperature. This monitored data will not be adequate to directly assess the heat-loss factor. However, together with billed quantities and information from the audit, this information can be used in the same way as BEP calculations are performed. The difference is that specific energy is more accurately assessed and temperatures are more precise.

The degree-day concept has been presented above and is repeated here. For the heating season

$$HDD = \sum_{t_{start}}^{t_{end}} (\theta_{HDD\_base} - \theta_e) \tag{3.32}$$

while for the cooling season

$$CDD = \sum_{t_{start}}^{t_{end}} (\theta_e - \theta_{CDD\_base}) \tag{3.33}$$

The internal temperatures θ_{HDD_base} and θ_{CDD_base} are base indoor temperatures for the period considered, and normalization with respect to the external climate often uses default values for normalizing space-heating and space-cooling energies. In equation form, these are

$$Q_{spaceheating}^{N\,delivered} = Q_{spaceheating}^{delivered} \cdot \frac{HDD^{N}}{HDD} \tag{3.34}$$

$$Q_{spacecooling}^{N\,delivered} = Q_{spacecooling}^{delivered} \cdot \frac{CDD^{N}}{CDD} \tag{3.35}$$

The base temperatures are default values and roughly correspond to the temperatures that would have been found if heat gains had been omitted in the building, yet the heating system delivered the same quantity.

From the MEP point of view, the mean temperature of the internal environment θ_i is assessed. Although the monitoring period is short, the assumption is that this internal climate is constant throughout the season. Therefore, a calculation is made for the actual and the reference years, such that, for the heating season

$$HDD = \sum_{t_{start}}^{t_{end}} (\theta_i - \theta_e) \tag{3.36}$$

while for the cooling season

$$CDD = \sum_{t_{start}}^{t_{end}} (\theta_e - \theta_i) \qquad (3.37)$$

Note especially that θ_i is completely different from the base temperatures used earlier. Whereas the base temperature refers to the effect of space heating/cooling only, θ_i also involves heat gains.

At a national level, the normal mean temperature of the internal environment θ_i^N is prescribed. The degree-days of the reference (normal) heating and cooling periods are computed according to

$$HDD_{ie}^N = \sum_{t_{start}}^{t_{end}} (\theta_i^N - \theta_e^N) \qquad (3.38)$$

for the normal heating season, and

$$CDD_{ie}^N = \sum_{t_{start}}^{t_{end}} (\theta_e^N - \theta_i^N) \qquad (3.39)$$

for the normal cooling season. The subscript ie denotes that normalization takes into consideration both reference internal and external climates.

The normalization calculations are then made with the equations as follows:

$$Q_{spaceheating}^{N\,ie\,delivered} = (Q_{spaceheating}^{delivered}) \cdot \frac{HDD_{ie}^N}{HDD}$$
$$+ (Q_{appliances}^{delivered} + Q_{metabolic} + Q_{solarheat}) \cdot \left(\frac{HDD_{ie}^N}{HDD - 1}\right)$$
$$(3.40)$$

$$Q_{spacecooling}^{N\,ie\,delivered} = (Q_{spacecooling}^{delivered}) \cdot \frac{CDD_{ie}^N}{CDD}$$
$$+ (Q_{appliances}^{delivered} + Q_{metabolic} + Q_{solarheat}) \cdot \left(1 - \frac{CDD_{ie}^N}{CDD}\right)$$
$$(3.41)$$

The modified utilization factor method

The modified utilization factor method does not explicitly make use of internal temperatures. Calculations are based on the set-point temperature of the main heating system. When normalization is only with respect to the external climate, the actual set-point temperature θ_{spt} is used.

For normalization with respect to internal climate, the changes will appear in the 'normal' (reference) set-point temperature, here called θ_{spt}^N. The numerical value for the reference set-point temperature is prescribed at a national level.

The six working steps of the methodology are to be followed, with the substitution of one variable. The changes are listed below.

$$Q_{heat\,losses}^{N\,ei\_spt} = Q_{heat\,losses}^{at\_spt} \cdot \frac{(\theta_{spt}^N - \theta_e^N)}{(\theta_{spt} - \theta_e)} \qquad (3.42)$$

$$\gamma^{N\,ei} = \frac{Q_{appliances}^{delivered} + Q_{metabolic} + Q_{solarheat}^N}{Q_{heat\,losses}^{N\,ei\_stp}} \qquad (3.43)$$

The normalized utilization factor is calculated using the normalized gain–loss ratio, here expressed as

$$\eta_{UF}^{N\,ei} = \frac{1 - (\gamma^{N\,ei})^a}{1 - (\gamma^{N\,ei})^{a+1}} \quad \text{if } (\gamma^{N\,ei}) \neq 1 \qquad (3.44)$$

$$\eta_{UF}^{N\,ei} = \frac{a}{a+1} \quad \text{if } (\gamma^{N\,ei}) = 1 \qquad (3.45)$$

The normalized space-heating energy is calculated from the set of variables as

$$Q_{spaceheating}^{N\,ei\,delivered} = Q_{heat\,losses}^{N\,ei\,at\_spt} - \eta_{UF}^{N\,ei} (Q_{appliances}^{delivered} + Q_{metabolic} + Q_{solarheat}^N) \qquad (3.46)$$

If the set-point temperature cannot be assessed during the audit or during the monitoring campaign (for example, by analysing temperature profiles during the night), the alternative is to use the lowest recorded temperature during the measurement period as the set-point temperature.

Normalization with heat-loss factor methods

The three procedures of normalization in the previous sections involve reference internal and external climates and are applied to MEP calculations. Primarily the short-term monitoring, together with the use of bills, allows the procedures to be applied.

For the extended monitoring period, which is primarily dedicated to determining the heat-loss factor of the building, another alternative can be employed. This is by using simulations. The experimental heat-loss factor, together with reference internal and external climates, is used for determining the normalized space-heating energy (and space-cooling energy).

Below, brief information is given on two methods that can be utilized.

THE ENERGY SIGNATURE METHOD

The Energy Signature (ES) method provides an experimental value for the heat-loss factor, H. With a set of prescribed internal and external temperatures, solar radiation and optionally internal heat gains, the space-heating energy for the normalized conditions can be calculated. A general formulation of a static model will be such that

$$Q_{spaceheating}^{N\,ei} = H(\theta_i^N - \theta_e^N) \cdot t - \sum A_{s,n} \cdot q_{g,n}^N - \Phi_{int\,gains}^N \cdot t \qquad (3.47)$$

where

t = the simulation time step (for example 1 month)

θ_i^N and θ_e^N = environment temperatures for the reference year

$A_{s,n}$ = solar apertures in direction n

$q_{g,n}^N$ = global radiation onto a surface in direction n for the reference year

$\Phi_{int\,gains}^N$ = the power of internal gains, possibly with normal values

This method is similar to the MUF method. Whereas the MUF method implicitly calculates a heat-loss factor for each data period considered, the ES method makes use of the mean heat-loss factor for all periods.

THE UA AND gA METHOD

The UA and gA method, developed by Somogyi,[5] makes use of an RC network and a Monte-Carlo approach to identify the UA value of the envelope. The calculation involves a simulation of the building subjected to a reference external climate and a fixed set of internal conditions, see Wouters and Loncour[6] and the literature review in Chapter 1.

NORMALIZATION WITH RESPECT TO BUILDING SIZE

A way of normalizing energy consumption to obtain a value that is independent of the size of the building is by dividing the energy use by the heated floor area or the volume enclosed by the building. The heated floor area of the building is denoted by A. The energy consumption, either global or specific, is generally defined as

$$q = \frac{Q}{A} \quad (\text{kWh/m}^2 \cdot \text{year}) \tag{3.48}$$

The definition of the heated or living floor area has a large impact on the magnitude of the area-specific energy requirement.

DEFINITION OF THE FLOOR AREA

Relating energy use to floor area is a common concept. However, national traditions and applications have different definitions for the size of floor area. These can vary from being the external or internal measure of the perimeter of the building envelope; they may also account for or exclude internal partition constructions, staircases, etc; and they may take into account whether or not spaces of the exposed floor are heated/cooled.

Within the framework of this project, and the application of the methodology to buildings, the proposed definition of the 'heated floor area' is defined as in the following.

The heated floor area is the sum of areas within the building envelope, based on the interior measures along the building envelope, which are exposed to living spaces that have a temperature that exceeds θ_{heated} (see definition below). The floor area includes the areas covered by partition constructions.

The definition of θ_{heated} is such that

$$\theta_{heated} = \bar{\theta}_e + 0.7 \cdot (\theta_{spt} - \bar{\theta}_e) \tag{3.49}$$

where

$\bar{\theta}_e = $ the average temperature of the external environment during the heating season (°C)

$\theta_{spt} = $ is the set-point temperature of the main heating system (°C). If this value is not available, default values for indoor temperatures determined at a national level can be utilized.

The following areas are *not* considered to be a part of the heated floor area:

- rooms lacking heating units and having a space temperature less than θ_{heated}
- heated garages if these are outside the building envelope
- passively heated garages if the space temperature is less than θ_{heated}
- external enclosed volumes with passive heating or occasional heating, such as glazed balconies
- sun-courts without heating units or where the temperature is less than the value θ_{heated}
- the lower horizontal areas of heated or non-heated crawl spaces
- cellars without active heating systems
- saunas that are used less than twice a week
- storage rooms without heating units and with closed entrances.

The following areas are considered to be a part of the heated floor area, although the space to which the floor is exposed to may not be equipped with a heating unit:

- spaces that lack heating units but are completely surrounded by heated spaces (completely internal spaces)
- spaces that lack heating units but are equipped with ventilation inlet terminals (for exhaust air)
- Halls and corridors with doors that are constantly open to adjacent heated spaces
- open staircases (area = projection onto each horizontal plane).

Area-specific space-heating energy

Energy for space heating is conveniently related to floor area:

$$q_{spaceheating}^{superscript} = \frac{Q_{spaceheating}^{superscript}}{A} \quad (\text{kWh/m}^2 \cdot \text{year}) \tag{3.50}$$

where the superscript can be :

- *delivered* for area-specific delivered space-heating energy
- *N delivered* for area-specific normalized delivered space-heating energy
- *supplied* for area-specific supplied space-heating energy
- *N supplied* for area-specific normalized supplied space-heating energy.

Equivalent area-specific space-cooling energy

Space cooling is a non-linear phenomenon and, when utilized, it is seldom that the whole building is cooled for a long period of time. Moreover, the fact that only some of the rooms are cooled, if these are cooled at all, gives rise to problems when defining what floor area is actively being cooled. Therefore, the same definition is used as for heated floor area, but is called the *equivalent* area-specific space-cooling energy

$$q_{spacecooling}^{superscript} = \frac{Q_{spacecooling}^{superscript}}{A} \quad (\text{kWh/m}^2 \cdot \text{year}) \tag{3.51}$$

where the superscript can be:

- *delivered* for area-specific delivered space-cooling energy
- *N delivered* for area-specific normalized delivered space-cooling energy
- *supplied* for area-specific supplied space-cooling energy
- *N supplied* for area-specific normalized supplied space-cooling energy.

Area-specific global energy

Annual area-specific global energy is the sum of specific energies, divided by the heated floor area. Expressed in equation form, the actual area-specific global energy use is formulated such that

$$q_{global}^{superscript} =$$

$$\frac{Q_{spaceheating}^{superscript} + Q_{spacecooling}^{superscript} + Q_{appliances}^{superscript} + Q_{hotwater}^{superscript} + Q_{external}^{superscript}}{A}$$

(3.52)

where the superscript can be:

- *delivered* for area-specific delivered global energy
- *N delivered* for area-specific normalized delivered global energy
- *supplied* for area-specific supplied global energy

- *N supplied* for area-specific normalized supplied global energy.

REFERENCES

1. EN ISO 13790: 2002, *Thermal Performance of Buildings – Calculation of Energy Use for Space Heating.*
2. NEN 2916, Energiprestatie van utilitels-gebouwen/Energy performance of non-residential buildings. Bepalingsmethode, Nederlands Normalisatie-instituut NEN, Delf, The Netherlands (in Dutch).
3. EUROCLASS, 2001, Final report of EUROCLASS: Development of an European Methodology to Experimentally Assess and Clasify Existing Residential Buildings Based on Their Actual Energy Consumption, Contract No. XVII/4.1031/Z/99–330, EU-SAVE programe.
4. Ministerio de Fomento e IDEA, 1999, *Fundamentos Técnicos de la Calificación Energética de Viviendas.*
5. Somogyi Z, 1998, *In Situ Evaluation of the Thermal Characteristics of Building Components and Buildings including Comparison with Predicted Performances.* PhD thesis UCL, Louvain-La-Neuve.
6. Wouters P and Loncour X, 2001, *In Situ Identification of UA and gA-value: An Overview of Possibilities and Difficulties.* Save Euroclass report, Department of Building Physics, Indoor Climate and Building Services, Belgian Building Research Institute, Brussels.

CHAPTER 4

The Euroclass method – description of the software

SERVANDO ÁLVAREZ, ANTONIO BLANCO, JUAN ANTONIO SANZ AND FRANCISCO J SÁNCHEZ

Engineering School, DIE-Grupo Termotecnia, University of Serville.
Camino de los descubrimientos s/n. Sevilla, Spain

THE RATING METHODOLOGY

Any rating procedure is a comparison scheme that makes it possible to give a score to a certain building (the rated building). It is based on three issues:

- *the variable of performance* or the set of variables that are going to be compared
- *the comparison scenario*, that is, the group of buildings (or the group of values) that are going to provide the distribution of the variable of performance, creating the framework of comparison
- *the rating score*, which includes the criteria and the limits that give the score when the variable of performance is compared in the comparison scenario.

A common framework for a rating methodology has been developed within the framework of the Euroclass project. It can be widely applied to the different methods (deterministic or experimental) of characterization of the energy consumption of residential buildings. At the same time, it can be adapted to national peculiarities and energy policy criteria. The framework is based on the use of the relative frequency distribution curves for the different end-uses of the energy. For every similarity level these curves are assumed to be independent of the characterization method.

The rated variables are: *total supplied energy* in kWh/m^2 and *total delivered energy* in kWh/m^2 for heating and/or cooling purposes.

These two variables can mainly be obtained either from the Billed Energy Protocol (BEP) or from the Monitored Energy Protocol (MEP), explained in Chapter 2, although a deterministic approach would also be acceptable. For every case, the rating procedure uses the results specific to the approach used.

In order to compare buildings within a common context, two successive normalizations are performed: normalization with respect to the climatic conditions and normalization with respect to the operating conditions.

A simple Excel-based program has been prepared to implement the rating methodology. Figure 4.1 shows the output screen of the software mentioned.

EUROCLASS: DEVELOPMENT OF A EUROPEAN METHODOLOGY TO EXPERIMENTALLY ASSESS AND CLASSIFY EXISTING RESIDENTIAL BUILDINGS BASED ON THEIR ACTUAL ENERGY CONSUMPTION CONTRACT NO. XVII/4.1031/Z/99–330

EUROTARGET

A COMPUTER TOOL TO IMPLEMENT THE EUROCLASS RATING PROCEDURE

Description of the software

This software implements the rating procedure developed for the Billed Energy Protocol (BEP) and for the Monitored Energy Protocol (MEP), explained in Chapters 2 and 3. The software makes it possible to introduce general data for the building characteristics and the site, as well as from one of the above-mentioned protocols. Each of these protocols provides useful information for carrying out a rating test of the building either in the cooling season or in the heating season. The program provides a rating for the building in a specific comparison scenario.

The tool has been developed in Excel workbook format. It includes four sheets. Two sheets contain the input–output variables, which ask for the required data and rating choices and present the results. Two further sheets include a database with information to be completed with country-dependent requirements, and a calculation sheet.

Input and output sheets

The input sheet is named 'Main data 1'. This sheet includes two windows in which climatic and either heating or cooling data are provided, for BEP or MEP. The output sheet is called 'Results' and includes the comparison scenario and

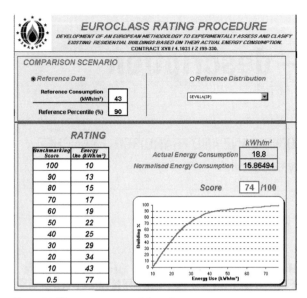

Figure 4.1 The output screen of the software program

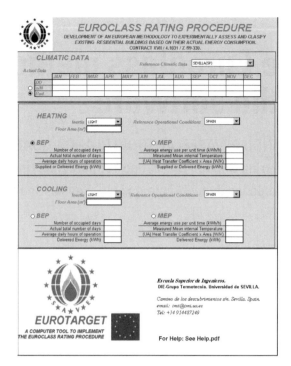

Figure 4.2 The input screen of the software program

the rating results. Actual and normalized energy consumptions, as well as the building score, are presented in this sheet. A more detailed description can be found in the following sections.

The database

A third sheet, named 'Update Database', includes the database, which will probably be edited to provide national data for the rating. The data are stored in tables that correspond to the windows in the input and output sheets.

CLIMATIC DATA SETS

Data for about 300 locations can be stored. The name of the set is required (city (country)) and is shown in the combo set. 48 values are located after the name, four-monthly sets for degree-days, n/N, total over horizontal surface radiation (kWh/m^2) and outside temperature.

INERTIA RATIOS

Three inertia ratios have been taken into account, light, mean and heavy. These values are calculated as $e^{-1/\tau}$ where τ is the main time constant of the building in hours.

HEATING REFERENCE CONDITIONS

This set of values comprises the reference conditions against which internal operation normalization will be carried out. About 100 values can be filled in.

COOLING REFERENCE CONDITIONS

These are similar to the heating reference conditions, but for the cooling-dominant season.

REFERENCE RATING DISTRIBUTION

About 100 different distributions can be included. Data required are the 10th percentiles for every distribution set. The format for the values is as a percentage.

DISTRIBUTION CALCULATION SET OF VALUES FOR REFERENCES

The ratio of the 90/10 percentiles, λ (lambda), and the 10% inflection percentile.

THE CALCULATIONS SHEET

The fourth (hidden) sheet is named 'Calculations' and must not be edited. It is the basis for the calculations that are used to carry out the rating.

Input and output screens

The input and output screens are shown in Figures 4.2 and 4.3.

Description of variables

DESCRIPTION OF THE INPUT

Figures 4.4 to 4.9 show areas of the input screen with annotations describing the different areas.

DESCRIPTION OF THE OUTPUT

Figures 4.10 to 4.12 show areas of the output screen with annotations describing the different areas.

DESCRIPTION OF THE DATABASE

Figures 4.13 to 4.15 show areas of the database screen with annotations describing the different areas.

Example

Figures 4.16 and 4.17 show the input and output screens for an example application.

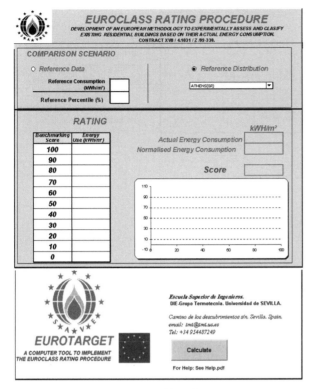

Figure 4.4 An empty input screen of the software program

Figure 4.3 An empty output screen of the software program

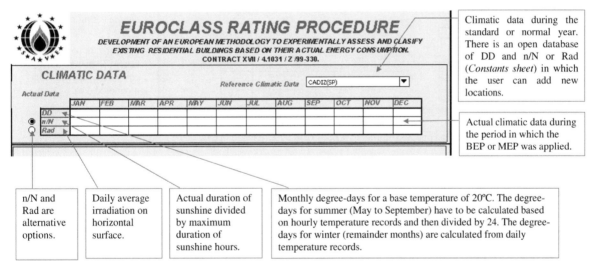

Climatic data during the standard or normal year. There is an open database of DD and n/N or Rad (*Constants sheet*) in which the user can add new locations.

Actual climatic data during the period in which the BEP or MEP was applied.

n/N and Rad are alternative options.

Daily average irradiation on horizontal surface.

Actual duration of sunshine divided by maximum duration of sunshine hours.

Monthly degree-days for a base temperature of 20°C. The degree-days for summer (May to September) have to be calculated based on hourly temperature records and then divided by 24. The degree-days for winter (remainder months) are calculated from daily temperature records.

Figure 4.5 Area of the input screen. Climatic data

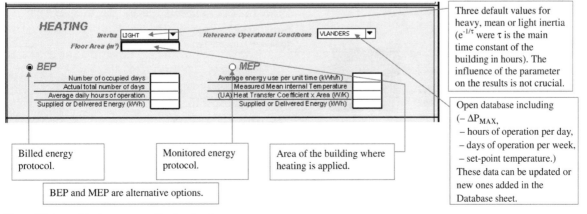

Three default values for heavy, mean or light inertia ($e^{-1/\tau}$ were τ is the main time constant of the building in hours). The influence of the parameter on the results is not crucial.

Open database including
($- \Delta P_{MAX}$,
– hours of operation per day,
– days of operation per week,
– set-point temperature.)
These data can be updated or new ones added in the Database sheet.

Billed energy protocol.

Monitored energy protocol.

Area of the building where heating is applied.

BEP and MEP are alternative options.

Figure 4.6 Area of the input screen. Heating data 1

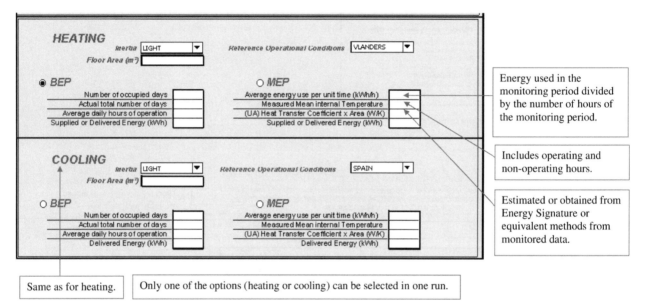

Number of days in which the building was occupied during the heating season (obtained from the questionnaire to the user).

Total number of days of the heating season.

Supplied energy: Energy that is actively supplied to the conversion systems of the residence. Supplied energy is often referred to as billed energy.
Delivered energy: Energy that is delivered into the residential living space through energy conversion systems.

Self-explanatory. (obtained from the questionnaire).

Figure 4.7 Area of the input screen. Heating data 2

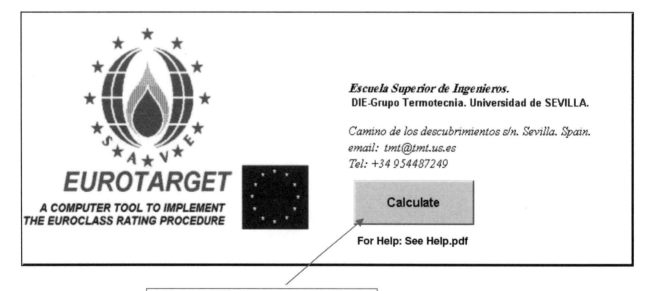

Energy used in the monitoring period divided by the number of hours of the monitoring period.

Includes operating and non-operating hours.

Estimated or obtained from Energy Signature or equivalent methods from monitored data.

Same as for heating.

Only one of the options (heating or cooling) can be selected in one run.

Figure 4.8 Area of the input screen. Heating and cooling data

Escuela Superior de Ingenieros.
DIE-Grupo Termotecnia. Universidad de SEVILLA.

Camino de los descubrimientos s/n. Sevilla. Spain.
email: tmt@tmt.us.es
Tel: +34 954487249

Calculate

For Help: See Help.pdf

EUROTARGET
A COMPUTER TOOL TO IMPLEMENT THE EUROCLASS RATING PROCEDURE

Once all the data has been introduced, click on Calculate button to run the program.

Figure 4.9 Area of the input screen. Calculate and Help buttons

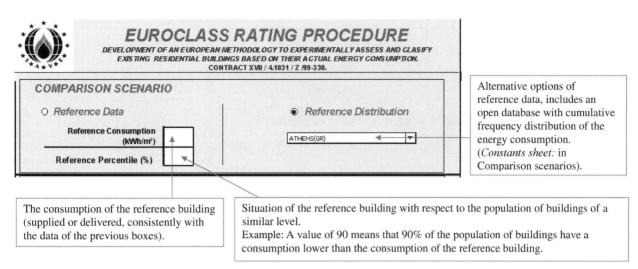

Alternative options of reference data, includes an open database with cumulative frequency distribution of the energy consumption. (*Constants sheet:* in Comparison scenarios).

The consumption of the reference building (supplied or delivered, consistently with the data of the previous boxes).

Situation of the reference building with respect to the population of buildings of a similar level.
Example: A value of 90 means that 90% of the population of buildings have a consumption lower than the consumption of the reference building.

Figure 4.10 Area of the output screen. Comparison scenario

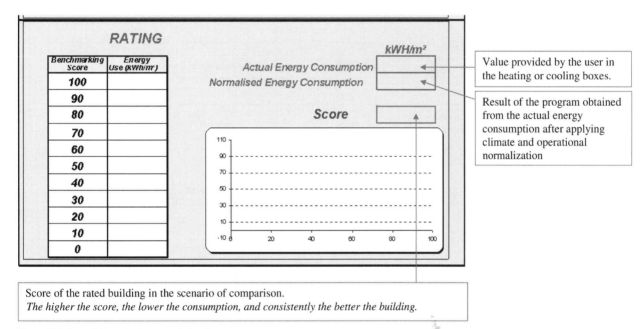

Value provided by the user in the heating or cooling boxes.

Result of the program obtained from the actual energy consumption after applying climate and operational normalization

Score of the rated building in the scenario of comparison.
The higher the score, the lower the consumption, and consistently the better the building.

Figure 4.11 Area of the output screen. Rating 1

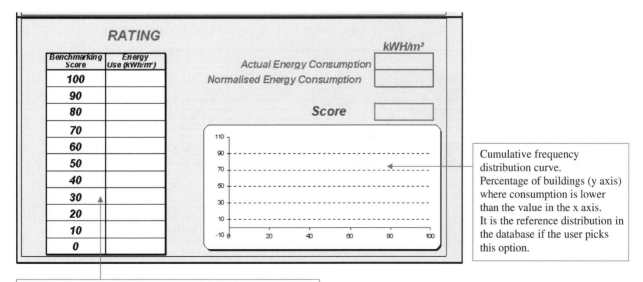

Cumulative frequency distribution curve.
Percentage of buildings (y axis) where consumption is lower than the value in the x axis.
It is the reference distribution in the database if the user picks this option.

Indicates the score obtained for different energy consumption.

Figure 4.12 Area of the output screen. Rating 2

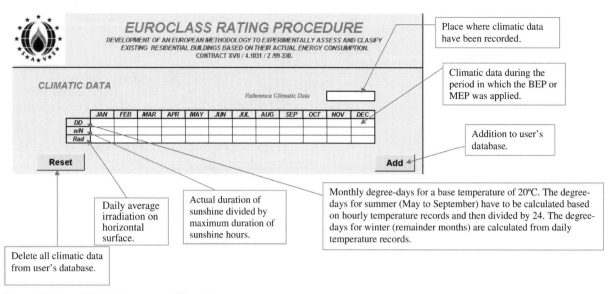

Figure 4.13 Area of the database screen. Climatic data

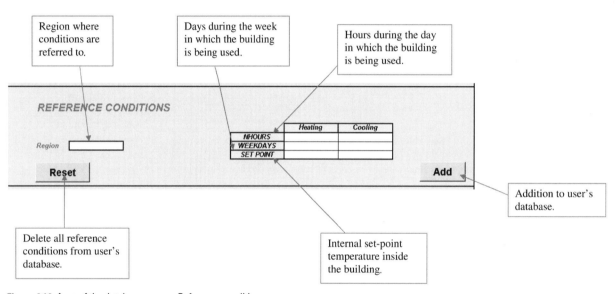

Figure 4.14 Area of the database screen. Reference conditions

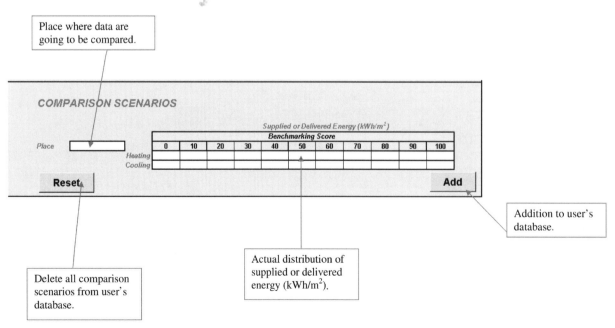

Figure 4.15 Area of the database screen. Comparison scenarios

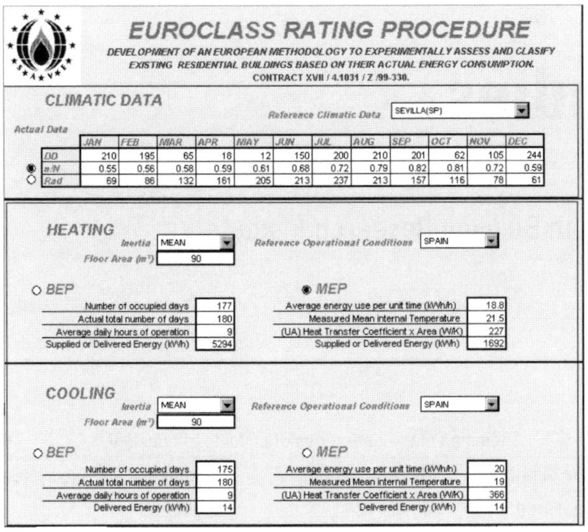

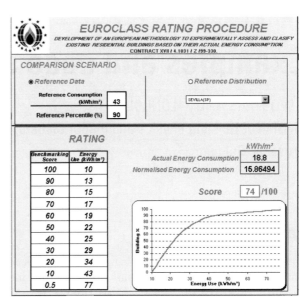

Figure 4.16 Input screen of the example

Figure 4.17 Output screen of the example

CHAPTER 5

Examples and case studies

P. WOUTERS AND X. LONCOUR

Division of Building Physics and Indoor Climate, Belgian Building Research Institute, Brussels, Belgium

INTRODUCTION

The methodology in the preceding chapters has been applied in four countries (Belgium, Greece, Spain and Sweden). This chapter describes the results of the procedure, the problems encountered and the solutions envisaged to solve them.

This chapter is divided into two parts, the first part deals with the results of the Billed Energy Protocol procedure (BEP) and the second part with the results of the Monitored Energy Protocol (MEP).

The two protocols have each been applied in the four considered countries. The application of the methodology, the way to implement the monitoring, the choice of the normalization techniques and the calculations have been implemented at a national level. The authors who have applied the different protocols and implemented the calculations are listed at the end of the chapter.

BUILDING SELECTION – CHARACTERISTICS OF THE BUILDING ANALYSED

Selection criteria

Different major criteria have been considered in the selection of the houses:

- The way the energy consumption and the energy bills are determined: energy bills based on *estimation* (estimated energy consumption) or on a *measurement* (measured energy consumption).
- The possibility of distinguishing within the energy consumption (of gas for instance), the part used for heating/cooling from the part used for other applications (cooking, hot tap-water production). If it is possible to completely distinguish the energy consumption for heating/cooling, we speak about *single-use of type of fuel*; if not, we speak about *multiple-use of type of fuel*.
- The final parameter to consider is the presence or absence of a cooling system.

Following these criteria, the next matrix was set up that defines the type of buildings to select. Four major categories of building were defined; these four categories can be split into two further sub-categories (single-use/multiple-use); see Table 5.1.

Table 5.1 Definition of the categories of building to consider

Type of bills	Heating system	Heating and cooling system
Measured	Single-use	Single-use
	Multiple-use	Multiple-use
Estimated	Single-use	Single-use
	Multiple-use	Multiple-use

Figure 5.1 The four countries considered

The type of buildings selected in the four countries

The diversity of climate encountered in the four countries considered (Figure 5.1) makes it possible to test the methodology in very different situations with different billing systems.

Buildings have been further subdivided according to the following considerations:

- In Belgium two buildings have been analysed for the heating consumption on the basis of bills (estimated energy consumption): one building in the 'single-use' category, the other in the 'multiple-use' category.
- In Sweden two buildings have been analysed for the heating consumption on the basis of bills (measured energy consumption): one building in the 'single-use' category, the other in the 'multiple-use' category.
- In each of Greece and Spain two buildings have been analysed for the heating consumption and cooling consumption on the basis of bills (measured energy consumption): one building in the 'single-use' category, the other in the 'multiple-use' category.

Table 5.2 Distribution of the type of test between the countries considered

Type of bills	Heating system	Heating and cooling system
Measured	Sweden	Spain and Greece
Estimated	Belgium	

Table 5.3 Normalization techniques chosen in the four countries considered

Country	Heating	Cooling
Belgium	Degree-days	Not relevant
Greece	Degree-days/ Climate Severity Index	Climate Severity Index
Spain	Climate Severity Index	Climate Severity Index
Sweden	Modified utilization factor method	Not relevant

A total of eight buildings have been analysed. The whole measurement programme is summarized in the Table 5.2.

Procedures and normalization techniques applied

In the four countries, both procedures (BEP and MEP) have been applied. The choice of the normalization technique to be applied was left to be decided at a national level. As described in Chapter 3, three techniques are available to normalize the heating/cooling consumption:

- degree-days method (DDM)
- modified utilization factor method (MUFM)
- Climate Severity Index technique (CSI).

For a description of these three normalization techniques, their advantages and limitations please see Chapter 3.

The techniques have been applied in the different countries as shown in Table 5.3. The normalized results presented in the next section have been established on the basis of the normalization techniques specified in this table.

RESULTS OF THE BILLED ENERGY PROTOCOL PROCEDURE
Introduction

The BEP procedure has already been extensively described in Chapter 2. We recall some aspects here. The Billed Energy Protocol (BEP) is based on:

- The collection of energy bills.
- A set of data collected on site. Data are collected on the type of building, the type of heating/cooling/ ventilation/hot tap-water systems, the appliances, the efficiency of the boilers, the set-point temperatures, the behaviour of the users, etc. All these data make it possible to produce a better analysis of the energy bills.
- The climatic data measured in meteorological stations.

No special monitoring is realized within the scope of this protocol.

Belgian buildings
THE BELGIAN BILLING SYSTEM

In Belgium, for most of the time, measured bills (related to gas, electricity or water consumption) are available once a year. The bills related to the supply of heating oil are available at each delivery. In the majority of cases, only one meter is installed in the building and all the consumption is measured with this meter. Multi-use of fuel is common. In the multi-family houses, several systems exist with separate or common measurements.

Monthly bills are sent based on the energy consumption of the previous year.

ESTIMATION OF THE UNKNOWN PARAMETERS OF THE ENERGY CONSUMPTION

Since multi-use of fuel is the most common system encountered in Belgium, it will often be necessary to *estimate* different parts of the global energy consumption (for instance energy consumption for hot water or for cooking). Each method of *estimation* is based on a number of assumptions.

Energy consumption for hot water

A calculation method applied in Belgium allows the assessment of the average hot-water consumption.[1] This method is generally used to dimension solar hot-water systems. The parameters playing a role in the calculated value are the number of rooms, the localization and the standing of the dwelling. This method is used in the scope of these calculations.

The method gives a figure for hot-water consumption. In order to determine the energy consumption, it is necessary to make assumptions about temperatures and about the efficiency of the production system.

Energy consumption for cooking

In some cases, multi-use of fuel occurs with the energy consumption for cooking (for instance in the case of gas). It is therefore necessary to estimate this energy consumption. Information coming from reference [2] makes it possible to assess this consumption.

NORMALIZATION – DEGREE-DAYS CORRECTION

Several choices can be made for normalizing the energy consumption with respect to the climate (outdoor temperature). Within the scope of the BEP protocol, the normalization of the energy consumption has been realized by using the degree-days 15/15 concept:

- The first 15°C means that the heating system works when the outside temperature is lower than or equal to 15°C. This first value makes it possible to determine the length of the heating season.
- The second 15°C means that the heating system is shut down when the indoor temperature reach 15°C. This second value determines the ΔT to take into consideration.

In practice, the duration of the heating season is different in each building; houses that are better insulated with a higher glazing area will have a shorter heating season than other buildings. The choice of a standardized calculation procedure allows the procedure to be applied systematically.

Other choices could be made to normalize with respect to the climate (degree-days 15/17). The impact of this decision is estimated below.

THE NORMAL BELGIAN CLIMATE

Figure 5.2 shows the monthly average outdoor temperature for a normal year in Uccle, while Figure 5.3 summarizes the monthly solar energy received during a normal year on a horizontal surface.

BUILDING NO. 1

General information

The first Belgian building selected is a four-storey single-family dwelling built in 1956. The house is located in the centre of Brussels (Figure 5.4). The total heated floor area is 273 m². Apart from the roof, the house is not insulated. There is both single and double glazing.

Currently, the building is occupied by two persons and is unoccupied for about 20 weeks per year. The occupancy pattern has radically changed during recent years. This is clearly shown in the results of the BEP procedure.

The heating system is a central heating system with gas, and the hot-water production is provided by the same boiler as the heating system. The system is used for hot-water production in the winter as well as in the summer. The regulation of the heating system occurs via a programmable thermostat situated in the living room. Only one meter for the gas consumption is installed in the house. An assessment of the hot-water consumption was necessary. Measured bills are available once per year.

One meter each is also installed for the water and electricity consumption. Measured annual bills are available.

No cooling system is installed.

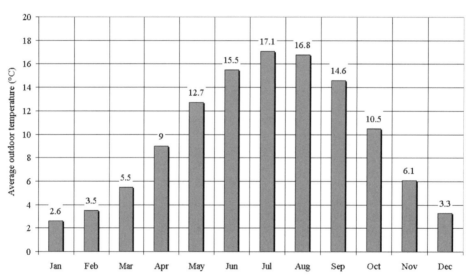

Figure 5.2 Monthly average outdoor temperature – normal year in Uccle (Brussels)

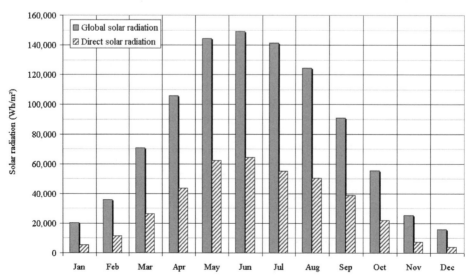

Figure 5.3 Monthly data for solar radiation – normal year in Uccle (Brussels)

Figure 5.4 Belgian building no. 1: street façade (left); garden façade (right)

Available data – calculation and results

A set of energy bills is available for this house. The procedure has been applied to the data for the four last years. Table 5.4 shows the meteorological data for these four years.

For this house, the bills are delivered in June between the two heating seasons. The data included in the energy bills are summarized in Table 5.5. In Table 5.6 other information necessary to assess the specific energy consumption is summarized.

Table 5.7 summarizes the measured and assessed supplied energy for the specific uses.

The following set of assumptions is used to assess the hot-water consumption:

- hot-water consumption (m^3) – assessed
- water temperature as supplied (average temperature assessed at 10°C) and temperature of the hot water delivered (set-point temperature equal to 57°C)
- system efficiency (see Table 5.6).

Based on these values, it is possible to calculate the supplied actual and normalized energy consumptions (Table 5.8). The normalization occurs here according to the 15/15 degree-days approach.

The results are very stable. We can clearly see the change of occupancy pattern in the house during the last year considered. These values are shown on Figure 5.5. An error band has been calculated by applying a variation of ±3°C on the temperature.

The numerical values of the error band are presented in Table 5.9. These values are given in both absolute and relative terms. It can be seen that the accuracy seems to be very good (maximum 6% deviation compared to the calculated value).

Table 5.4 Belgian climatic data

Year	Average outdoor temperature during the winter (°C)	Degree-days 15/15
Normal year	6.9	2,033
1996–1997	6.7	2,067
1997–1998	8.1	1,658
1998–1999	7.5	1,824
1999–2000	8.2	1,730

The area-specific results are presented in Table 5.10. The same error band in the relative values is valid for these results. Figure 5.6 shows these area-specific values.

The results for the delivered energy are presented in Table 5.11.

The fact that the bills are delivered in June makes it easy to use the data to calculate the energy consumption. The degree-days for only one winter are included in the bill.

A different normalization procedure could be used, but the stability of the results would be the same.

A major characteristic of the methodology can be seen here. This is that no normalization is carried out for the occupancy pattern, which means that the results presented are very dependent on the occupancy. New occupants would mean that a new analysis of the energy consumption would be required.

BUILDING NO. 2
General information

The second building considered in Belgium is a detached building, which was built in 1985 and is situated in the country. The building is shown in Figure 5.7. The total

Table 5.5 Supplied energy – based on bills

Supplied energy	Number of days between bills	Number of occupants	Number of weeks away	Gas (kWh)	Electricity (kWh)
Estimated (E) or measured (M)				M	M
Conversion factor				10.28 kWh/m^3	1 kWh/kWh
17 June 1996–16 June 1997	364	4	3	39,392	5,964
16 June 1997–12 June 1998	361	4	3	34,074	5,360
12 June 1998–17 June 1999	370	3	5	35,353	5,378
17 June 1999–28 June 2000	377	2	20	24,666	3,970

Table 5.6 Form table 2: List of energy conversion systems, their efficiencies and how these were assessed

Energy conversion system	Fuel type	Winter efficiency	Summer efficiency	Estimated (E), measured (M) or calculated (C)
Household appliances + lighting	Electricity	100%	100%	E
Heating system gas	Gas	60%	–	E
Hot-water system – combined with the heating system	Gas	40%	40%	E

Table 5.7 Supplied energy (in kWh)

		Energy consumption – gas	Hot-water consumption	Cooking consumption	Q_{heat}	Energy consumption – electricity	Q_{tot}
Measured (M), estimated (E) or calculated (C)		M	E	E	C	M	C
Year	Number of days						
1996–1997	364	39,392	5,372	923	33,097	5,980	45,372
1997–1998	361	34,074	5,416	931	27,726	5,419	39,493
1998–1999	370	35,353	5,069	871	29,413	5,305	40,658
1999–2000	377	24,666	3,387	582	20,697	3,844	28,509

Table 5.8 Supplied actual and normalized energy consumptions

Year	$Q_{heat, sup}$ (kWh/year)	$Q_{heat, sup, normal}$ (kWh/year)	$Q_{tot, sup}$ (kWh/year)	$Q_{tot, sup, normal}$ (kWh/year)
1996–1997	33,097	32,561	45,372	44,836
1997–1998	27,726	33,996	39,493	45,763
1998–1999	29,413	32,797	40,658	44,042
1999–2000	20,697	24,333	28,509	32,146

Table 5.10 Supplied area-specific actual and normalized energy consumption

Year	$qA_{heat, sup}$ (kWh/m^2 year)	$qA_{heat, sup, normal}$ (kWh/m^2 year)	$qA_{tot, sup}$ (kWh/m^2 year)	$qA_{tot, sup, normal}$ (kWh/m^2 year)
1996–1997	121	119	166	164
1997–1998	102	125	145	168
1998–1999	108	120	149	161
1999–2000	76	89	104	118

Table 5.9 Supplied actual and normalized energy consumption – error band

Year	$Q_{heat, sup, normal}$ (kWh/year)	$Q_{tot, sup, normal}$ (kWh/year)
1996–1997	32,561 < Q < 33,067 (Delta max 1.6%)	44,836 < Q < 45,342 (Delta max 1.1%)
1997–1998	33,168 < Q < 35,954 (Delta max 5.8%)	45,763 < Q < 47,720 (Delta max 4.3%)
1998–1999	32,546 < Q < 33,023 (Delta max 0.9%)	43,791 < Q < 44,268 (Delta max 0.6%)
1999–2000	24,333 < Q < 24,872 (Delta max 2.6%)	32,146 < Q < 32,685 (Delta max 1.7%)

Table 5.11 Delivered normalized energy (in kWh)

Year	$Q_{heat, N, del}$	$q_{heat, N, del}$	$Q_{tot, N, del}$	$q_{tot, N, del}$
1996–1997	19,536	71.6	27,665	101.3
1997–1998	20,398	74.7	27,984	102.5
1998–1999	19,678	72.1	27,011	98.9
1999–2000	14,600	53.5	19,798	72.5

heated floor area is 276 m^2. The house is unoccupied for about two weeks per year.

The heating system is a central heating system with gas. The production of hot water is realized with a separate boiler and a solar system. Two gas meters are installed in the house; the first general meter measures the total gas consumption of the building and the bills are established on the basis of readings from this meter. The second meter measures the gas consumption for hot water (the readings are taken manually and there has been no systematic record in recent years). This kind of situation is exceptional in the Belgian context. We are have here the use of a single fuel (except for the energy consumption for cooking, which is marginal).

Available data – calculation and results

The energy bills are available for recent years. The bills are delivered once a year in February. It is necessary to split

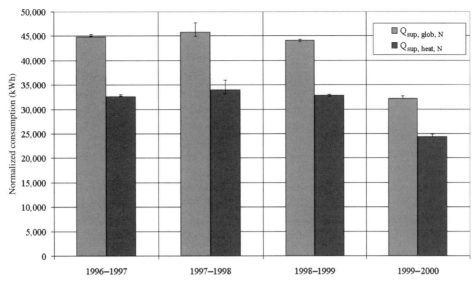

Figure 5.5 Belgian building no. 1: Evolution of the normalized total and heating energy consumption over four years

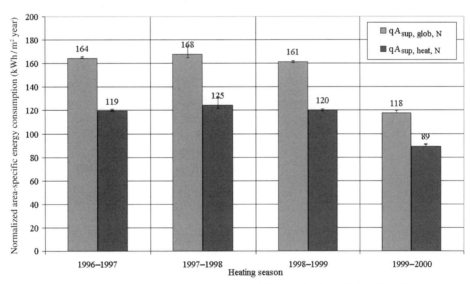

Figure 5.6 Belgian building no. 1: Evolution of the area-specific normalized supplied total and heating energy consumption over four years

Figure 5.7 Belgian building no. 2: Pictures of the house

the energy consumption over the two winters covered by the bill to establish the energy consumption of the building. Unfortunately, manual readings of the second gas meter are not available. Within the BEP procedure, multiple use of fuel was considered.

The meteorological data are available and have already been presented in Table 5.4. In Table 5.12, the calculation carried out to split the consumption between the two winters is summarized, while the data taken from the bills are summarized in the Table 5.13 and the efficiency of the different systems in Table 5.14.

The main information coming from the bills is summarized in Table 5.15. An assessment of the contribution of the solar system to the energy consumption for hot water has been made (50%). On this basis, the energy consumption for heating has been calculated. Table 5.16

shows the energy consumption normalized with respect to the energy consumption of the appliances, together with the normalized area-specific energy consumption.

The energy consumption is relatively constant over the three years. The specific consumptions are rather low, which can be explained by the good level of insulation of the building and the building design developed to increase the solar gains.

The results are presented with an error band in Table 5.17. These results have been calculated by applying a variation of ±3°C in the DD correction. The values are given in both absolute and in relative terms. The results are represented graphically in Figures 5.8 and 5.9.

Based on the efficiency values given in Table 5.14, Table 5.18 gives the values of the delivered energy, while the area-specific results are presented in Table 5.19.

Table 5.12 Calculations carried out to split the energy consumption between the two winters

Year	Number of days between the two bills	Number of days in winter 1	Number of days in winter 2	Degree-days in winter 1	Degree-days in winter 2
1997–1998	360	101	125	564	1,218
1998–1999	374	86	144	579	1,439
1999–2000	357	75	136	503	963

Table 5.13 Supplied energy, based on bills

Supplied energy	Number of days between bills	Number of occupants	Number of weeks away	Gas (kWh)	Electricity (kWh)
Estimated (E) or measured (M)				M	M
Conversion factor				10.32 kWh/m³	1 kWh/kWh
17 February 1997–12 February 1998	360	4	2	12,484	3,750
12 February 1998–21 February 1999	374	4	2	16,156	3,780
21 February 1999–20 February 2000	357	4	2	12,504	3,849

Table 5.14 Form table 2: List of energy conversion systems, their efficiencies and how these were assessed

Energy conversion system	Fuel type	Winter efficiency	Summer efficiency	Estimated (E), measured (M) or calculated (C)
Household appliances + lighting	Electricity	100%	100%	E
Heating system gas	Gas	75%	–	M
Hot-water heating system – gas (solar pre-heated hot water)	Gas	75%	75%	E

Table 5.15 Supplied energy (in kWh)

		Energy consumption – gas	Hot-water consumption	Contribution of solar system	Cooking consumption	Q_{heat}	Energy consumption – electricity	Q_{tot}
Measured (M), estimated (E) or calculated (C)		M	E	E	E	C	M	C
Year	Number of days			% solar system				
1997–1998	360	12,484	2,133	50%	1,360	8,991	3,750	16,234
1998–1999	374	16,156	2,216	50%	1,413	12,527	3,780	19,936
1999–2000	357	12,504	2,115	50%	1,349	9,040	3,849	16,353

Table 5.16 Normalized and area-specific supplied energy

Year	$Q_{N, heat, sup}$ (kWh/year)	$q_{N, heat, sup}$ (kWh/m$^2 \cdot$ year)	$Q_{N, tot, sup}$ (kWh/year)	$q_{N, tot, sup}$ (kWh/m$^2 \cdot$ year)
1997–1998	10,258	37.2	17,602	63.7
1998–1999	12,621	45.7	19,851	71.9
1999–2000	12,542	45.4	20,019	72.5

Table 5.17 Supplied actual and normalized energy consumption – error band

Year	$Q_{heat, sup, normal}$ (kWh/year)	$Q_{tot, sup, normal}$ (kWh/year)
1997–1998	$10,258 < Q < 10,701$ (Delta max 4%)	$17,602 < Q < 18,044$ (Delta max 2.5%)
1998–1999	$12,486 < Q < 13,193$ (Delta max 4.5%)	$17,716 < Q < 20,424$ (Delta max 2.9%)
1999–2000	$12,287 < Q < 14,268$ (Delta max 13.8%)	$19,764 < Q < 21,744$ (Delta max 8.6%)

Swedish buildings

THE SWEDISH BILLING SYSTEM

The Swedish billing system is usually based on monthly bills. The billed quantities are either measured values or predicted values based on earlier measurements. Readings are made (or should be made) at least once per year. However, there is a change taking place, where more sophisticated monitoring equipment is being installed to make fuel debits based on real monthly consumption. The fast development of on-line systems or a tele-infrastructure makes it possible for these types of services to be introduced.

As a result of national energy policies after the oil crisis in the 1970s, many single-family dwellings (47%) are equipped with electrical heating. The wider deployment of the experimental on-line collection protocol should make it possible to monitor the multipurpose use of electricity. Monitoring equipment that makes use of a building's central distribution unit is convenient. Although measurement of

electrical entities is relative simple, there may be problems with non-billed fuels. In rural areas, it is common for biomass fuels to be gathered locally and combusted for the purpose of heating, in stoves, fireplaces and boilers. The only records are usually what the owner/end-user remembers. Furthermore, since the mid 1990s, the government has issued subsidies for installing biomass stoves.

Multi-family buildings are commonly heated by means of district heating. A major problem here is that most buildings are not equipped with devices that monitor the consumption within individual apartments. It is very common for all energy costs, apart from those for electricity for domestic appliances and lighting, to be shared by residential tenants/occupants on the basis of floor area. With an infrastructure that lacks meters for each apartment, MEP is likely to be the protocol that is applied for these buildings.

CLIMATE DATA

The Swedish Meteorological and Hydrological Institute (SMHI) collects climate data. With 372 sites across Sweden, the availability of climate data is high. However, not all stations provide data on solar radiation and the sampling frequency may differ depending on the site. For the purpose of the experimental on-line protocol, apart from solar radiation, temperature measurements and degree-day data are available on a month-wise basis.

For Sweden, there is a so-called 'normal year' for various regions. The normal year is the average of 30 years, represented by data for 1961–1990. Global solar data is available for the normal year, but not for the diffuse and direct radiation onto a normal (perpendicular) surface. For this reason, data has been used for the period 1983–1998.[3] It was not until 1983 that diffuse and direct radiation were continuously measured at several sites.

Normalization of external climate is based on the normal year. The degree-day data provided by SMHI assumes a base temperature of 17°C. By default, the heating season

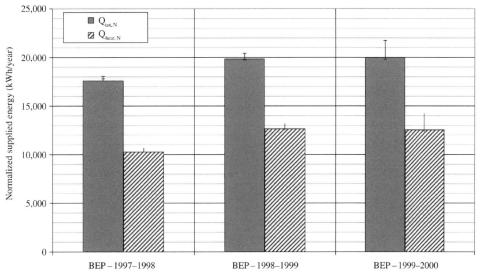

Figure 5.8 Belgian building no. 2: Evolution of the normalized total and heating energy consumption over three years

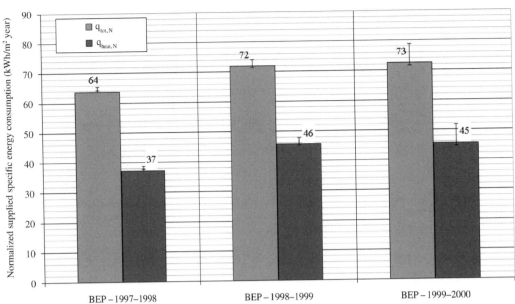

Figure 5.9 Belgian building no. 2: Evolution of the normalized area-specific total and heating energy consumption over three years

Table 5.18 Delivered normalized energy (in kWh)

	Hot-water consumption	Cooking consumption	$Q_{heat, N}$	Energy consumption – electricity	$Q_{tot, N}$
Measured (M), estimated (E) or calculated (C)	E	E	C	M	C
Year					
1997–1998	3,374	1,379	7,694	3,802	16,248
1998–1999	3,374	1,379	9,465	3,689	17,907
1999–2000	3,374	1,379	9,407	3,935	18,094

Table 5.19 Delivered normalized energy (in kWh)

Year	$Q_{heat\_N\_del}$	$q_{heat\_N\_del}$	$Q_{tot, N, del}$	$q_{tot\_N\_del}$
1997–1998	7,694	27.9	16,248	58.8
1998–1999	9,465	34.3	17,907	64.9
1999–2000	9,407	34.1	18,094	65.5

is considered to last from 15 September until 15 May; the summer period starts when the average daily temperature is higher than 10°C.

Figure 5.10 shows the monthly average outdoor temperature in Sweden for the normal year, while Table 5.20 gives some representative data.

BUILDING NO. 1

General information

The apartment analysed is situated in a semi-detached two-family house built in 1982, subject to Swedish building code SBN 80. The building contains two apartments in all. The building is relatively well insulated and the mechanical ventilation is equipped with a heat exchanger. Situated next to a hill in a suburban area, the building is generally shaded from the sun. Figure 5.11 shows pictures of the building.

Until 1993, the only fuel type was electricity (multipurpose). The installation of a wood fan stove in 1993

changed the fuel usage. From then on, the electrical water-based heating system was used as the base heating system, with a constant set-point temperature of 17°C. The heating system is operated all year round, while the fan stove is used as a complementary unit. Wood is gathered from the grounds of the summer cottage, so there are no bills for this fuel. Nor are there any detailed records of wood consumption. The occupier has therefore estimated the use of wood on a seasonal (annual) basis, partly based on knowledge of the global energy use prior to the installation of the fan stove and partly on the basis of the volume of the shed where the wood is stored. Prior to the installation of the stove, electricity bills indicated a usage of about 20,000 kWh/year but this was for four persons in the apartment.

Supplied electricity is monitored on an hourly basis. On a daily basis, the supplier collects data at midnight over a telephone connection. Monitored values are actually listed on the bill, which is sent to the customer at the start of the following month. Within the energy audit of BEP, the monthly summed values have been used.

Formally, it is not evident that BEP can be applied in this case because of the uncertainty about the amount of wood concerned. However, within the framework of the development of the methodology, it was interesting to evaluate this case using BEP. MEP application will provide

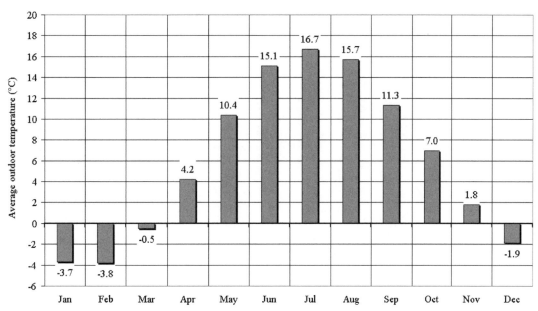

Figure 5.10 Monthly average outdoor temperature in Stockholm, Sweden for a normal year

Table 5.20 Climate data variables for mean temperature, direct solar radiation onto a normal surface, diffuse solar radiation and the global solar radiation onto a horizontal surface for a normal year

	Actual year	Normal year	Actual year			Normal year		
	Mean temperature	Mean temperature	Direct solar radiation	Diffuse solar radiation	Global solar radiation	Direct solar radiation	Diffuse solar radiation	Global solar radiation
Month	(°C)	(°C)	(Wh/m²)	(Wh/m²)	(Wh/m²)	(Wh/m²)	(Wh/m²)	(Wh/m²)
December 1999	−1.6	−1.9	16,484	4,860	5,900	13,600	4,500	6,000
January 2000	−1.2	−3.7	33,285	7,661	12,080	18,300	7,400	10,000
February 2000	−0.1	−3.8	48,482	16,842	28,040	38,000	16,400	25,600
March 2000	1.6	−0.5	138,989	31,982	80,640	68,400	34,400	59,800
April 2000	6.7	4.2	96,270	51,972	95,770	103,700	54,500	104,400
May 2000	12.2	10.4	150,266	75,099	153,190	169,000	68,500	160,900
June 2000	14.3	15.1	137,121	87,221	161,920	146,800	76,000	158,800
July 2000	16.3	16.7	102,765	69,320	124,400	156,700	72,500	160,000
August 2000	15.9	15.7	131,930*	34,670*	99,050*	121,800	61,400	123,100
September 2000	11.1	11.3	150,235	36,016	95,540	92,700	37,400	75,200
October 2000	10.5	7.0	18,314	20,890	26,030	49,800	21,300	35,600
November 2000	6.8	1.8	3,286	5,965	6,520	21,300	8,400	12,000
December 2000	2.6	−1.9	1,551	4,315	4,480	13,600	4,500	6,000

*Not available; reference year 1971 data used for August.

more detail on the daily consumption of wood, and should therefore be more reliable. How large will the differences in results be? Another problem that this type of consumption poses is: should it be considered to be climate-dependent or climate-independent? Wood is only used during the winter, but the increase in consumption as the weather gets colder is not known.

Water consumption is not explicitly available. The official meter registers the total consumption of 18 households and the costs are shared among them on an equal basis. The occupier installed a private water meter in September 1993, but has not made any recordings since then. The first registered reading was made during the audit.

Monthly climate data was obtained for the year 2000 from a climate station in the region of Bromma (some 3 km away) and solar radiation from a station situated some 10 km away.

Available data – calculation and results

The information that is available from the bills and entered in the form tables is summarized in Tables 5.21 to 5.24, with the system efficiencies of the different systems and how these were assessed given in Table 5.22.

Table 5.24 lists the periods for which estimations on space heating have been made. These correspond to each month starting at December 1999 and ending at November 2000. The periods correspond to those of the electrical bills, but the wood consumption has been distributed among the months of the heating season in proportion to the temperature difference between the internal and the external climates. Derived values for delivered specific energy are entered in the next column. The MUF method has been used for normalizing the space heating. Data is entered on a month-wise basis in the preceding columns,

Figure 5.11 Swedish building no. 1: Pictures of the house. The street façade is on the left and the garden façade on the right

Table 5.21 Form table 1: Supplied energy

Supplied energy	Electricity	Wood
Estimated (E) or measured (M)	M	E by occupant
Conversion factor	1.00 kWh/kWh	1,500 kWh/m³
December 1999	1,874	
January 2000	1,847	
February 2000	1,459	
March 2000	1,426	
April 2000	1,040	
May 2000	688	
June 2000	509	
July 2000	474	
August 2000	438	
September 2000	640	
October 2000	769	
November 2000	963	
Heating season December 1999–November 2000	–	3,000
Sum (kWh/year)	12,127	3,000

first for the actual year and then for the normal year. The data needed are the dry-bulb temperature for the external environment, direct solar radiation onto the normal plane and diffuse solar radiation onto the horizontal plane.

For the building under examination, the nominal time constant was estimated to be 30 hours. Although the set-point temperature of the main heating system (the water-based radiator system) is 17°C, wood is burnt almost every day during the heating season. For this reason, the average temperature (18–20°C) given by the occupant has been used, corresponding to approximately 19°C. The values

from the processing of data are listed in the column headed normal space heating in Table 5.24. In the last column, a parameter analysis gives the minimum value for the delivered space heating to be 10,023 kWh. This was obtained for a time constant corresponding to 125% of the nominal time constant, a set-point temperature that is 19−3°C and solar apertures 50% larger than the assessed values. The maximum value was obtained for a time constant corresponding to 75% of the nominal time constant, a set-point temperature that is 19 + 3°C and solar apertures 50% less than the assessed values. Note that the set-point temperature in this calculation is 16°C, which is lower than the set-point temperature of the main heating system.

The final results of the procedures are shown in Tables 5.25 and 5.26.

Remarks

Although supplied electricity is measured in detail (hourly values available), the use of non-billed wood gives rise to problems when allocations of specific energy are to be calculated. The approximate estimation of wood consumption gives rise to quite unreliable results. This is one of the reasons that either the quantity of non-billed fuels should be verified by a third party or MEP should be applied. The distribution of wood use over the heating season was in this case assumed to be climate dependent. This assumption may be erroneous.

The normalization technique used was the MUF method. Although the set-point temperature of the main heating system was known, the use of wood burning with the purpose of heating the building makes the latter set-point temperature void. The real set-point temperature, which

Table 5.22 Form table 2: List of energy conversion systems, their efficiencies and how these were assessed

Energy conversion system	Fuel type	Winter efficiency	Summer efficiency	Estimated (E), measured (M) or calculated (C)
Household appliances + lighting	Electricity	1.00	1.00	E
Electrical boiler for space heating and hot water, 10.5 kW (1982)	Electricity	1.00	1.00	E
Fan stove, wood manual feeding (installed 1993)	Wood	0.85	0.85	E

Table 5.23 Form table 3: Allocation of delivered energy, on an energy conversion system level, for each of the specific energies

Method	Footnote 3		Footnote 2	Footnote 1	
Electrical conversion systems	Space heating	Space cooling	Appliances and lighting	Tap hot water	External appliances/spaces
December 1999	1,317		380	177	0
January 2000	1,290		380	177	0
February 2000	956		344	159	0
March 2000	869		380	177	0
April 2000	501		368	171	0
May 2000	131		380	177	0
June 2000	0		338	171	0
July 2000	0		297	177	0
August 2000	0		262	177	0
September 2000	101		368	171	0
October 2000	212		380	177	0
November 2000	424		368	171	0
Sum	5,801		4,248	2,079	0
Method	**Footnote 4**				
Wood fan stove	Space heating	Space cooling	Appliances and lighting	Tap hot water	External appliances/spaces
December 1999	379				
January 2000	479				
February 2000	450				
March 2000	405				
April 2000	271				
May 2000	126				
June 2000	0				
July 2000	0				
August 2000	0				
September 2000	0				
October 2000	171				
November 2000	268				
Sum	2,550				
Sum (all) kWh/year	8,351	0	4,248	2,079	0

1. Electricity delivered for tap-water heating is estimated. The energy use is estimated using a 'standard' value in Sweden, corresponding to 1,000 kWh/person per year.

2. A base load is estimated on the basis of energy use during the summer period. The mean value of electricity supplied during the period May–September is used, giving a load of 748.6 W for the winter season (some 18 kWh/day). From this value, the mean power for tap-water heating is subtracted. For the summer months, appliance energy is obtained by subtracting hot-water energy from delivered electricity.

3. The space-heating energy is estimated as the total delivered electricity minus the electricity for tap water (footnote 1) and for electrical appliances (footnote 2).

4. The total amount of wood was estimated by the occupant to be 2 m^3. The corresponding energy was distributed over the winter months by weighting according to the temperature difference between indoors and outdoors.

must be considered to be subjective (when the occupant feels the need for an increased internal temperature) cannot be read and will vary over the day.

Solar irradiation was on average 7% of the total delivered heat during the heating season. A minimum of around 1% was found in December. Since the solar radiation is very small, the concept of degree-day normalization (DD) can be applied. In Table 5.27, the new normalization parameters have been entered. The base temperature in Sweden is, by default, 17°C.

The results from application of the DD method give a value that is 3% less than the value obtained from the MUF method. A difference of 312 kWh is noted. The value given by DD lies within the fuzzy band of MUF.

BUILDING NO. 2

General information

Pictures of the building being analysed are given in Figure 5.12. The apartment being considered is situated in the middle of the roof of the building.

The apartment is situated in one of three buildings within the estate. The estate is owned by a co-operative-living society, and each occupant indirectly owns the apartment they live in. Built in 1998 under Swedish building code BBR94, the buildings are equipped with individual metering of each apartment; this is not the usual method of billing within Swedish co-operative-living societies. Traditionally, occupants pay for the electricity used within the apartment,

Table 5.24 Form table 4: Display of chosen processing periods, actual delivered space-heating energy for the periods, numerical values of influencing variables, and normalized space-heating energy. The bottom two rows indicate the band (minimum and maximum) values that are obtained with prescribed conditions according to the normalization methods. The method used for normalization was MUF

Period	Space heating	Actual exterior temperature	Actual direct solar radiation	Actual diffuse solar radiation	Normal exterior temperature	Normal direct solar radiation	Normal diffuse solar radiation	Normal space heating
December 1999	1,766	−1.6	16,484	4,860	−1.9	13,600	4,500	1,802
January 2000	1,730	−1.2	33,285	7,661	−3.7	18,300	7,400	2,020
February 2000	1,372	−0.1	48,482	16,842	−3.8	38,000	16,400	1,740
March 2000	1,248	1.6	138,989	31,982	−0.5	68,400	34,400	1,523
April 2000	769	6.7	96,270	51,972	4.2	103,700	54,500	1,021
May 2000	279	12.2	150,266	75,099	10.4	169,000	68,500	454
June 2000	0	14.3	137,121	87,221	15.1	146,800	76,000	5
July 2000	0	16.3	102,765	69,320	16.7	156,700	72,500	4
August 2000	0	15.9	131,930*	34,670*	15.7	121,800	61,400	8
September 2000	101	11.1	150,235	36,016	11.3	92,700	37,400	107
October 2000	397	10.5	18,314	20,890	7.0	49,800	21,300	698
November 2000	690	6.8	3,286	5,965	1.8	21,300	8,400	1,121
Sum	8,351							10,503
							Minimum	10,023
							Maximum	12,274

*Not available; reference year 1971 data used for August.

Table 5.25 Delivered energy; normalized values were obtained with MUF

Year	Space heating (kWh/year)	Area-specific space heating (kWh/m² per year)	Space cooling (kWh/year)	Area-specific space cooling (kWh/m² per year)	Global (kWh/year)	Area-specific global (kWh/m² per year)
December 1999–November 2000	8,351	70.2	0	0	14,677	123.3
Normal	10,503	88.3	0	0	16,829	141.4

Table 5.26 Supplied energy; normalized values were obtained with MUF

Year	Space heating (kWh/year)	Area-specific space heating (kWh/m² per year)	Space cooling (kWh/year)	Area-specific space cooling (kWh/m² per year)	Global (kWh/year)	Area-specific global (kWh/m² per year)
December 1999–November 2000	8,801	73.9	0	0	15,127	127.1
Normal	11,069	93.0	0	0	17,395	146.2

Table 5.27 Form table 4: Display of chosen processing periods, actual delivered space-heating energy for these periods, numerical values of influencing variables, and normalized space-heating energy; the method used for normalization is DD

Period	Space heating	Actual degree-days	Normal degree-days	Normal space heating
December 1999	1,766	578	579	1,769
January 2000	1,730	565	632	1,935
February 2000	1,372	496	593	1,640
March 2000	1,248	478	547	1,428
April 2000	769	300	384	984
May 2000	279	59	148	700
June 2000	0	8	7	0
July 2000	0	0	1	0
August 2000	0	0	2	0
September 2000	101	153	123	81
October 2000	397	184	305	658
November 2000	690	307	443	995
Sum	8,351			10,191
			Minimum	–
			Maximum	–

Figure 5.12 Swedish building no. 2: Pictures of the building. The street façade is on the left and a side view on the right

and the rest of the energy bills are shared and included in the rent.

With this kind of co-operative-living society, there are several players involved in the assessment of data:

- The board of the society must agree that a third party can access the data. However, data for an individual household can only be given to a third party if the individual occupants agree to this. The individual occupant can refuse delivery of data even though the board has consented.
- The occupant cannot themselves provide the data without the permission of the board. However, the board can only be given permission to release data if a majority of the members have formally voted and consented to this.
- Only the board, and not the individual member, can grant the maintenance operator permission to release the data.
- In this case, there are two maintenance operators. One is the ordinary maintenance operator (who in this case was the building entrepreneur) and the other specifically deals with the monitoring and control system.

With this situation, it is evident that the entire classification process must have all the legal aspects cleared at an early stage.

The monitoring system was installed and run by Landis-Staefa for Siemens. Sampling frequencies can be made every 2 seconds, but the aim was to make monthly debits for energy use (for this reason the temperatures have not

been recorded) and these values are saved. The measured variables are:

- space heating that is dissipated by the apartment radiators
- consumption of cold tap water
- consumption of hot tap water
- electricity for household appliances
- space heating and electricity used in common spaces and for common appliances; this also includes bathroom towel heaters in the apartments (not individually measured).

Water-based system and radiators deliver heat into the apartment. The source is district heating, both for space heating and for tap-water heating. Electricity is used for household appliances and lighting only. Energy delivered for the purpose of tap-water heating is included in the debiting system and is calculated by multiplying the quantity consumed by the specific heat capacity of the water and a temperature increase of 55°C.

Use of energy for common spaces and appliances is debited on basis of the apartment size (floor area) as a proportion of the sum of the total floor area of all the apartments. This energy is used for common space heating (halls, corridors and storage rooms) and for lighting, elevators, fans and laundries (whether or not the occupants use these common laundries).

The apartment considered is situated on top of one of the buildings, rather like a villa placed on the roof. There are two occupants, one of whom is regularly at home. The heated floor area is 106 m^2. Although the apartment is equipped with a fireplace, this is never used.

Table 5.28 Form table 1: Supplied energy

Supplied energy	Electricity (appliances)	District heating 1	District heating 2	District + electricity
Estimated (E) or measured (M)	M	M	M	M
Conversion factor	1.00 kWh/kWh	1.00 kWh/kWh	1.00 kWh/kWh	1.00 kWh/kWh
January 2000	573	2,497	197	656
February 2000	503	2,135	169	589
March 2000	482	1,830	186	582
April 2000	410	949	137	463
May 2000	406	229	202	152
June 2000	369	66	175	230
July 2000	302	0	164	111
August 2000	300	14	164	171
September 2000	329	158	148	208
October 2000	423	575	180	319
November 2000	399	1,296	142	356
December 2000	490	1,760	208	447
Sum (kWh/year)	4,986	11,509	2,072	4,284

Table 5.29 Form table 2: List of energy conversion systems, their efficiencies and how these were assessed

Energy conversion system	Fuel type	Winter efficiency	Summer efficiency	Estimated (E), measured (M) or calculated (C)
Household appliances + lighting	Electricity (appliances)	1.00	1.00	E
District heating 1 + 2; 1 = space heating, 2 = hot water	District heating	1.00	1.00	E
District heating + electricity of external spaces and appliances	District heating + electricity	1.00	1.00	E

Available data – calculation and results

The information given in Tables 5.28–5.31 is available from the bills.

For the current building, the nominal time constant was estimated to be some 30 hours. The set-point temperatures are 21.5°C during daytime and 19°C at night. In the calculations, the mean temperature is used. The minimum value for delivered space heating is 14,200 kWh, obtained for the time constant corresponding to 125% of the nominal time constant, a set-point temperature that is the nominal less 3°C and solar apertures that are 50% larger than the assessed values. The maximum value of 16,067 kWh was obtained for the time constant corresponding to 75% of the nominal time constant, a set-point temperature that is the nominal plus 3°C and solar apertures that are 50% less than the assessed values.

Actual and normalized values are displayed in Tables 5.32 and 5.33.

Remarks

Three numerical values from the assessment are interesting. The first is the energy for heating tap water. In Sweden, the default value of 1,000 kWh/person per year is often used. This corresponds well with the current value of 2,072 kWh for two persons. Second, the use of appliances and lighting was 4,986 kWh (a normal value for Swedish circumstances). This is of the same order as energy use for external space heating and appliances, which corresponds to 4,284 kWh. For this case, external energy use may not be neglected.

Although the legal aspects were complex, as discussed above, the technical aspects of the billing system as such were

beneficial for BEP. Delivered energy is directly measured and recorded. The protocol is easy to apply within the framework of apartments with individual metering, and this is practised, for example, in all of Germany and Switzerland.

Although the monitoring system in this case is equipped with the option of recording the temperature where the central thermostats are installed, this option has not been used. If it had been, the quality of the data is such that the heat-loss factor could have been estimated within BEP. Since the set-point temperatures are known from the audit, these could be used as base temperatures.

As an optional service, estimation of the heat-loss factor of the building is possible if the quality of data from supplied energy bills is judged to be adequate. This was done for the considered apartment. Two methods have been tested:

- The sum of the delivered space-heating energy and the gains is assumed to equal the heat loss at a constant internal temperature. The temperature is assumed to be the mean value of the set-point temperatures, 21.5°C (daytime) and 19.0°C (night-time).
- The sum of the delivered space-heating energy and part of the gains, determined on the basis of the utilization factor according to the modified utilization factor (MUF) method, is assumed to equal the heat loss at the mean set-point temperature, as above. The time constant of the building was assumed to be 30 hours.

The mean rate of heat loss is calculated on a monthly basis, and regressed against the difference between the set-point temperature and the external temperature. All monthly data are included in the regression. The results are plotted in

Table 5.30 Form table 3: Allocation of delivered energy, on an energy conversion system level, to each of the specific energies

Method			Footnote 1		
Household appliances + lighting	Space heating	Space cooling	Appliances and lighting	Tap hot water	External appliances/spaces
January 2000			573		
February 2000			503		
March 2000			482		
April 2000			410		
May 2000			406		
June 2000			369		
July 2000			302		
August 2000			300		
September 2000			329		
October 2000			423		
November 2000			399		
December 2000			490		
Sum			4,986		

Method	Footnote 2			Footnote 3	
District heating 1 + 2	Space heating	Space cooling	Appliances and lighting	Tap hot water	External appliances/spaces
January 2000	2,497			197	
February 2000	2,135			169	
March 2000	1,830			186	
April 2000	949			137	
May 2000	229			202	
June 2000	66			175	
July 2000	0			164	
August 2000	14			164	
September 2000	158			148	
October 2000	575			180	
November 2000	1,296			142	
December 2000	1,760			208	
Sum	11,509			2,072	

Method					
District heating + Electricity	Space heating	Space cooling	Appliances and lighting	Tap hot water	External appliances/spaces
January 2000					656
February 2000					589
March 2000					582
April 2000					463
May 2000					152
June 2000					230
July 2000					111
August 2000					171
September 2000					208
October 2000					319
November 2000					356
December 2000					447
Sum					4,284
Sum (all) kWh/year	11,509	0	4,986	2,072	4,284

1. Electricity delivered to the apartment is measured and billed at the end of each month.

2. Delivered space heating is measured for the apartment and billed at the end of each month.

3. The quantity of tap water, hot and cold separately, is measured. The system operator calculates a temperature increase for hot water corresponding to 55°C times the specific heat capacity of water. This quantity is billed.

4. External energy use (measured) for common-space heating, lighting, laundries and operation of installations (fans, pumps and elevators).

Table 5.31 Form table 4: Display of chosen processing periods, actual space-heating energy for the periods, numerical values of the influencing variables, and normalized space-heating energy. The bottom two rows indicate the band (minimum and maximum) values that are obtained with the conditions prescribed according to the normalization method of MUF

Period	Space heating	Actual external temp	Actual direct solar	Actual diffuse solar	Normal external temperature	Normal direct solar	Normal diffuse solar	Normal space heating
January 2000	2,497	−1.2	33,285	7,661	−3.7	18,300	7,400	2,940
February 2000	2,135	−0.1	48,482	16,842	−3.8	38,000	16,400	2,722
March 2000	1,830	1.6	138,989	31,982	−0.5	68,400	34,400	2,299
April 2000	949	6.7	96,270	51,972	4.2	103,700	54,500	1,206
May 2000	229	12.2	150,266	75,099	10.4	169,000	68,500	325
June 2000	66	14.3	137,121	87,221	15.1	146,800	76,000	361
July 2000	0	16.3	102,765	69,320	16.7	156,700	72,500	41
August 2000	14	15.9	131,930*	34,670*	15.7	121,800	61,400	140
September 2000	158	11.1	150,235	36,016	11.3	92,700	37,400	183
October 2000	575	10.5	18,314	20,890	7.0	49,800	21,300	926
November 2000	1,296	6.8	3,286	5,965	1.8	21,300	8,400	1,931
December 2000	1,760	−1.6	16,484	4,860	−1.9	13,600	4,500	2,346
Sum	11,509							14,932
							Minimum	14,200
							Maximum	16,067

*Not available; reference year 1971 data used for August.

Table 5.32 Delivered energy. Normalized values were obtained with MUF

Year	Space heating kWh/year	Area-specific space heating kWh/m^2 · year	Space cooling kWh/year	Area-specific cooling kWh/m^2 · year	Global kWh/year	Area-specific global kWh/m^2 · year
2000	11,509	108.6	0	0	18,567	175.2
Normal	14,932	140.9	0	0	21,990	207.5

Table 5.33 Supplied energy. Normalized values were obtained with MUF

Year	Space heating kWh/year	Area-specific space heating kWh/m^2 · year	Space cooling kWh/year	Area-specific cooling kWh/m^2 · year	Global kWh/year	Area-specific global kWh/m^2 · year
2000	11,509	108.6	0	0	22,851	215.6
Normal	14,932	140.9	0	0	26,274	247.9

Figure 5.13. The conclusions are as follows:

- For large temperature differences (>10°C), the utilization factor is high, which in the calculations gives a good agreement between the methods. There is also good agreement for the heat-loss factor in these regions.
- For small temperature differences (<10°C), the MUF method gives a good extrapolation of the wide range of temperature difference.
- The results from the two methods diverge for small temperature differences (<10°C) because of the influence of the utilization factor. This region, <10°C, is defined in Sweden as the summer season, and the traditional ES method would exclude this data in heat-loss factor calculations. However, when the 'raw data' was inspected, only July was without delivery of space heating. Nevertheless, the delivered heat is small in the period May to September, accounting for the five leftmost points for each method.

The heat-loss factor was 186 W/K for the Energy Signature (ES) method using data for all months. From the MUF method, the corresponding figures were 232 W/K. Table 5.34 shows the result of variations in the time constant.

In the case when only space heating is regressed against the temperature difference (excluding May to August, when space-heating energy is negligible), the heat-loss factor is 223 W/K. This value is close to the one obtained for the 50-hour time constant.

Normalization by the degree-day method (DD) was tested to study deviations compared to the MUF method. Relevant values are shown in Table 5.35. The difference between the results of the two methods is 420 kWh. The DD method gives results 3% lower incomparison with the value obtained with MUF. Almost half of this deviation is because the MUF method gives a heating requirement during the summer months; see for example August. Furthermore, the solar radiation for August 2000 was not available.

Spanish buildings

INTRODUCTION

In Spain, the energy billing system includes natural gas, electricity and fuel-oil use. Electricity bills and consumption measurements are distributed every two months by the Spanish utilities to all residences. Usually, there is a single meter per household for all electricity uses. On some

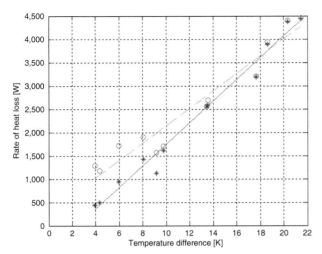

Figure 5.13 The heat-loss factor is determined from a linear regression of power versus temperature difference. The stars show the estimated heat loss at the set-point temperature as calculated by means of the MUF method; linear regression gave the solid line. The circles show the estimated heat loss that is equal to estimated delivered energy; linear regression gives the dashed line

Table 5.34 Dependency of the heat-loss factor on the time constant. The Energy Signature (ES) method is, independent of the time constant. In terms of MUF, the ES method can be considered as a special case of MUF where the utilization factor is constant with a value of 1

Method	MUF	ES
Time constant (h)	Heat-loss factor (W/K)	Heat-loss factor (W/K)
30	232	186
40	227	186
50	222	186

Table 5.35 Form table 4: Display of chosen processing periods, actual delivered space-heating energy for the periods, numerical values of influencing variables, and normalized space-heating energy. The method used for normalization is DD

Period	Space heating	Actual degree-days	Normal degree-days	Normal space heating
January 2000	2,497	565	632	2,793
February 2000	2,135	496	593	2,552
March 2000	1,830	478	547	2,094
April 2000	949	300	384	1,215
May 2000	229	59	148	574
June 2000	66	8	7	58
July 2000	0	0	1	–
August 2000	14	0	2	–
September 2000	158	153	123	127
October 2000	575	184	305	953
November 2000	1,296	307	443	1,870
December 2000	1,760	448	579	2,275
Sum	11,509			14,512
			Minimum	–
			Maximum	–

occasions there may be a second night-rate electricity meter.

Natural gas comes from the gas piping network. It is billed every two months with an indication of the actual consumption.

Diesel oil is billed with the purchase of fuel (once or twice in a year). For a multi-storey apartment building, the fuel bills are distributed every month by the building manager, together with the other common-use charges. In new apartment blocks the occupants of each apartment have the capability to turn on and off the heating according to their needs and each flat has its own metering device for billing purposes. Central heating systems are mostly of the hot water/radiator type. There is a central boiler room and a fuel tank.

Although there are no official reference meteorological years, there are degree-days and design conditions for sizing heating and cooling equipment. In forthcoming regulations, synthetic meteorological years are used for 50 locations. These years have been obtained from monthly average values of the climatic variables obtained for the period 1961–1990 as published by the World Meteorological Organization (WMO).

The two buildings analysed and monitored are situated in Seville. Tables 5.36 to 5.38 and Figures 5.14 and 5.15 give the relevant climatic variables.

Table 5.36 Heating degree-days at 20°C for the heating season of Seville

Reference	Year
January	276.2
February	210.6
March	170.8
November	155.3
December	261.7
Season	1,074.6

Table 5.37 Heating degree-days at 20°C for the cooling season of Seville

Reference	Year
June	50.1
July	105.7
August	105.4
September	59.8
Season	321

Table 5.38 Mean ambient temperatures for Seville

Reference	Year
January	10.7
February	11.9
March	14.0
April	16.0
May	19.6
June	23.5
July	26.9
August	26.8
September	24.4
October	19.5
November	14.3
December	11.1

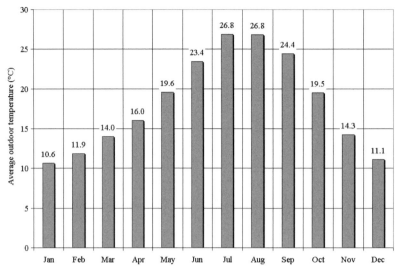

Figure 5.14 Monthly average outdoor temperature – Seville

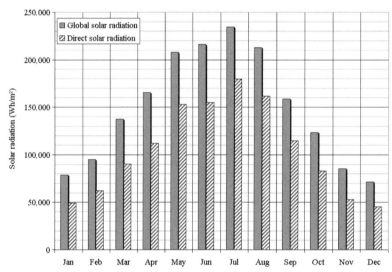

Figure 5.15 Normalized solar radiation in Seville

BUILDING NO. 1

General information

The building is a six-storey multi-family building constructed in 1998. It is located near the city centre of Seville (Figure 5.16). Each apartment has an air-conditioned floor area of $90\,m^2$ and a gross area of some $130\,m^2$. The exterior walls consist of $12\,cm$ of brick, $4\,cm$ of insulation (polystyrene), $4\,cm$ of hollow brick and an interior finishing of $1.5\,cm$ of plaster. The wall of the west façade is covered outside by $2\,cm$ of white stone. All the windows are double-glazed. The ratio of windows to walls is very high in the west façade.

At present, five people occupy the considered apartment throughout the year except for August. The occupancy pattern is that of a typical Spanish residence. Everybody goes out of the house early in the morning and comes home again to have lunch at two in the afternoon. Later the occupancy is more variable, but it could be said that half of the people go out and return during the evening. During the weekends, either Saturday or Sunday is spent away.

The air-conditioning system is separate for each apartment. It consists of an air-to-air heat pump for both heating and cooling. The domestic hot-water system is an independent boiler without storage. A thermostat in the living room controls the air temperature. Setback temperatures are selected during both winter and summer seasons.

There are individual meters for electricity, natural gas and water (but not for hot water). Electricity bills are received every two months; water and natural gas every three months.

Available data – calculation and results

The data are given in Tables 5.39 to 5.44.

Remarks

The length of the heating and cooling seasons has been provided by the owner.

The degree-days for each billed period, both for heating and cooling, have been calculated from hourly temperature

Figure 5.16 Spanish building no. 1: Pictures of the building. The east façade is on the left and the west façade on the right

Table 5.39 Form table 1: Supplied energy

Supplied energy	Natural gas	Electricity
Estimated (E) or measured (M)	M	M
Conversion factor	10.8 kWh/Nm³	1.00 kWh/kWh
October–November 1999	810	1,224
December 1999–January 2000	994	2,280
February–March 2000	940	1,240
April–May 2000	950	1,135
June–July 2000	583	1,474
August–September 2000	551	1,149
Sum (kWh/year)	4,828	8,502

records during the actual periods. The results are shown in Table 5.45.

BUILDING NO. 2

General information

The building is a two-storey single-family house built in 1977 (Figure 5.17). It is located near the city centre of Seville. The house has an air-conditioned floor area of 200 m² and a gross area of some 240 m². The exterior walls consist of 12 cm of brick, 4 cm of insulation (polystyrene) and 12 cm of brick. All the windows are single glazed. The ratio of windows to walls is very high in the south façade, but an overhang is available.

At present five people occupy the building throughout the whole year except for August. The occupancy pattern is that of a typical Spanish residential house. Everybody goes out of the house early in the morning and comes home again to have lunch at two in the afternoon. Later the occupancy is more variable, but it could be said that half of the people go and return during the evening. During summer weekends, the house is usually empty.

The air-conditioning system is centralized for the whole house. It consists of an air-to-air heat pump for both heating and cooling. The domestic hot-water system consists of three electrically heated tanks with a storage capacity of 80 litres each and a power of 1.5 kW. A thermostat in the living room controls the air temperature. Setback temperatures are selected during both winter and summer seasons.

There are individual meters for electricity and water (but not for hot water). Electricity bills are available every two months; water bills every three months.

Available data – calculation and results

The data are shown in Tables 5.46 to 5.51.

Remarks

The lengths of the heating and cooling season have been provided by the owner.

The degree-days for each billed period, both for heating and cooling, have been calculated from hourly temperature records during the actual periods.

The hot-water consumption has been estimated assuming 45 l/person day, a supply temperature of 45°C and a mains water temperature of 18°C.

Greek buildings

INTRODUCTION

In Greece the energy billing system is mainly concerned with electricity and fuel-oil use. Electricity bills and consumption measurements are distributed every three months to all residences by the Greek Public Power Corporation (PPC). Usually, there is a single meter per household for all electricity uses (hot water, cooking, air conditioning etc.). In some cases there may be a second night-rate electricity meter.

Diesel oil is the main fuel for central heating systems. For a single house the fuel bill comes with the purchase of fuel (once or twice in a year). For a multi-storey apartment building the fuel bills are distributed every month by the building manager, together with the other common-use charges (according to the floor area for older flats). In newer apartment blocks the occupants of each apartment have the option to turn on and off the heating depending on their needs and each flat has its own metering device for billing purposes.

Central heating systems are mostly of the hot water/radiator type. There is a central boiler room and a fuel tank. Some very new buildings that are going to use natural gas are connected directly to the gas-piping network.

Table 5.40 Form table 2: List of energy conversion systems, their efficiencies and how these were assessed

Energy conversion system	Fuel type	Winter efficiency	Summer efficiency	Estimated (E), measured (M) or calculated (C)
Household appliances and lighting	Electricity	1.00	1.00	E
Natural gas boiler for hot water	Natural gas	0.65	0.65	E
Split-unit heat pump for space heating	Electricity	COP 3.3	–	E
Split-unit heat pump for air conditioning	Electricity	–	COP 2.7	E

Table 5.41 Form table 3: Allocation of delivered energy, on an energy conversion system level, to each of the specific energies

Method				Footnote 1	
Natural gas boiler	Space heating	Space cooling	Appliances and lighting	Tap hot water	External appliances/spaces
October–November 1999				526	
December 1999–January 2000				645	
February–March 2000				610	
April–May 2000				617	
June–July 2000				378	
August–September 2000				358	
Sum				3,134	
Method	**Footnote 2**	**Footnote 3**	**Footnote 4**		
Electrical conversion systems	Space heating	Space cooling	Appliances and lighting	Tap hot water	External appliances/spaces
October–November 1999	177		1,047		
December 1999–January 2000	1,251		1,029		
February–March 2000	176		1,064		
April–May 2000			1,135		
June–July 2000		392	1,082		
August–September 2000		546	603		
Sum	1,604	938	5,960		
Total (kWh/year)	1,604	938	5,960	3,134	

1. The hot water is supplied by the natural gas boiler unit (at around 55°C). Mean yearly efficiency has been estimated as:

$$\sum W_c \, kg \times 4.18\,kJ/kg\,K \times (55 - 18)\,K/3{,}600\,kJ/kWh\,/Q_{supplied\ hot\ water}$$

where $W_c = 4$ persons $\times$ 50 litres/day per person $\times$ number of days in the period.

2. $Q_{heating} = Q_{electricity} - Q_{appliances}$ for each period.
3. $Q_{cooling} = Q_{electricity} - Q_{appliances}$ for each period.
4. Based on the intermediate periods (neither heating nor cooling) we can estimate a uniform daily consumption by appliances of 17.74 kWh and thus $Q_{appliances} \approx 17.74 \times$ number of days in the period.

Table 5.42 Actual supplied energy for heating and cooling

Year	Actual heating energy consumption (kWh/year)	Actual heating energy consumption (kWh/m² · year)	Actual cooling energy consumption (kWh/year)	Actual cooling energy consumption (kWh/m² · year)
October 1999–September 2000	1,604	17.82	938	10.42

Table 5.43 Actual delivered energy for heating and cooling

Year	Actual heating energy consumption (kWh/year)	Actual heating energy consumption (kWh/m² · year)	Actual cooling energy consumption (kWh/year)	Actual cooling energy consumption (kWh/m² · year)
Normalized	5,293	58.81	2,532	28.13

The Athens reference meteorological year has been derived from the records of five official meteorological stations over the last 25 years. Tables 5.52 to 5.54 give some of the relevant climatic variables.

CSI estimation from a correlation of degree-days and insolation hours in Greece is calculated as follows:

$$CSI = m_1 DD + m_2 (n/N) + m_3 DD^2 + m_4 n/N^2 + b$$

Table 5.44 Normalized delivered energy for heating and cooling

Normalization*	Normalized heating energy consumption (kWh/year)	Specific heating energy consumption (kWh/m² · year)	Normalized cooling energy consumption (kWh/year)	Specific cooling energy consumption (kWh/m² · year)
Normalized #1	7,786	86.51		
Normalized #2	7,818	86.86		
Normalized #3	6,886	76.51	1,846	20.51

*Three methods have been used in order to obtain the normalized results.

Normalized #1: Normalized heating energy consumption = Actual heating energy consumption $\times$ $DD_{normalized\ year}/DD_{actual\ year}$.

Normalized #2: (Normalized heating energy consumption $-$ Constant $\times$ winter solar radiation) = (Actual heating energy consumption $-$ Constant $\times$ winter solar radiation) $\times$ $DD_{normalized\ year}/DD_{actual\ year}$.

Normalized #3: Normalized heating energy consumption = Actual heating energy consumption $\times$ $CSI_{normalized\ year}/CSI_{actual\ year}$ and Normalized cooling energy consumption = Actual cooling energy consumption $\times$ $CSI_{normalized\ year}/CSI_{actual\ year}$.

Table 5.45 Heating and cooling degree-days

Period	DD_h	DD_c
3 October–1 December 1999	167	0
1 December 1999–4 February 2000	431	0
4 February–5 April 2000	161	0
5 April–8 June 2000	7	22
8 June–8 August 2000	0	317
8 August–3 October 2000	0	174

Table 5.46 Form table 1: Supplied energy

Supplied energy	Electricity
Estimated (E) or measured (M)	M
Conversion factor	1.00 kWh/kWh
January–February 1999	3,549
March–April 1999	1,450
May–June 1999	2,391
July–August 1999	2,250
September–October 1999	1,658
November–December 1999	3,580
Sum (kWh/year)	14,878

where n is the actual number of insolation hours for each period, and N is the maximum number of insolation hours for that period.

The coefficients given in Table 5.55 are valid for both winter and summer periods. The cooling degree-days at 20°C are given in Tables 5.56 and a comparison of simple degree-days and the results of the CSI method is given in Table 5.57.

From the above results we see that there is a considerable difference between the DD and CSI normalization ratios, especially during the cooling season. For this reason the CSI method was considered to be more suitable for Greece.

Figures 5.18 and 5.19 show the data relative to the reference year.

BUILDING NO. 1

General information

The building (Figure 5.20) considered is a two-storey detached building constructed in 1992 and comprising two separate apartments. Only one apartment (situated on the top floor) is considered within this procedure.

The total heated area is equal to 78 m². Two adults and two children occupy the apartment. A natural ventilation system is used (opening windows).

A central heating system is installed running on diesel oil. The distribution of heat is realized via a water-borne system and radiators. The apartment is also equipped with a reversible heat pump running on electricity (split-unit air conditioner 2.8 kW, rated power 1.4 kW – coefficient of performance COP $\approx$ 2.0), which is used for auxiliary heating and cooling, from 2 to 4 hours per day during hot summer days.

Hot-water production (approximately 45% of the total water consumption) is carried out with the main heating system during the winter and with a back-up electrical boiler during the summer.

Figure 5.17 Spanish building no. 2: Pictures of the building. The street façade is on the left and the garden façade on the right

Table 5.47 Form table 2: List of energy conversion systems, their efficiencies and how these were assessed

Energy conversion system	Fuel type	Winter efficiency	Summer efficiency	Estimated (E), measured (M) or calculated (C)
Household appliances and lighting	Electricity	1.00	1.00	E
Split-unit heat pump for space heating	Electricity	COP 3.3	–	E
Split-unit heat pump for air conditioning	Electricity	–	COP 2.7	E

Table 5.48 Form table 3: Allocation of delivered energy, on an energy conversion system level, to each of specific energies

Method	Footnote 1	Footnote 2	Footnote 3	Footnote 4	
Electrical conversion systems	Space heating	Space cooling	Appliances and lighting	Tap hot water	External appliances/spaces
January–February 1999	2,237		761	551	
March–April 1999			841	609	
May–June 1999		1,033	788	570	
July–August 1999		823	828	599	
September–October 1999		231	828	599	
November–December 1999	2,153		828	599	
Sum	4,390	2,087	4,873	3,528	
Total (kWh/year)	4,390	2,087	4,873	3,528	

1. $Q_{heating} = Q_{electricity} - Q_{appliances}$ for each period.

2. $Q_{cooling} = Q_{electricity} - Q_{appliances}$ for each period.

3. Based on the intermediate periods (neither heating nor cooling) we can estimate a uniform daily consumption by appliances of 13.35 kWh and thus $Q_{appliances} \approx 13.35 \times$ number of days in the period.

4. The hot water is supplied by the electrical heated tanks (at around 55°C). Mean yearly efficiency has been estimated as:

$$\sum W_c \, (kg) \times 4.18 \, (kJ/kg \, K) \times (55 - 18) \, (K)/3{,}600 \, (kJ/kWh)/Q_{supplied \, hot \, water}$$

where $W_c = 5$ persons $\times$ 50 litres/day per person $\times$ number of days in the period.

Table 5.49 Actual supplied energy for heating and cooling

Year	Actual heating energy consumption (kWh/year)	Actual heating energy consumption (kWh/m² · year)	Actual cooling energy consumption (kWh/year)	Actual cooling energy consumption (kWh/m² · year)
January–December 1999	4,390	19.95	2,087	9.49

Table 5.50 Actual delivered energy for heating and cooling

Year	Actual heating energy consumption (kWh/year)	Actual heating energy consumption (kWh/m² · year)	Actual cooling energy consumption (kWh/year)	Actual cooling energy consumption (kWh/m² · year)
January–December 1999	14,487	65.85	5,634.9	25.61

Table 5.51 Normalized delivered energy for heating and cooling

Normalization*	Normalized heating energy consumption (kWh/year)	Specific heating energy consumption (kWh/m² · year)	Normalized cooling energy consumption (kWh/year)	Specific cooling energy consumption (kWh/m² · year)
Normalized #1	21,308	96.86		
Normalized #2	21,340	97.00		
Normalized #3	18,847	85.67	8,127	36.94

*Three methods have been used in order to obtain the normalized results.

Normalized #1: Normalized heating energy consumption = Actual heating energy consumption $\times$ DD$_{normalized \, year}$/DD$_{actual \, year}$.

Normalized #2: (Normalized heating energy consumption – Constant $\times$ winter solar radiation) = (Actual heating energy consumption – Constant $\times$ winter solar radiation) $\times$ DD$_{normalized \, year}$/DD$_{actual \, year}$.

Normalized #3: Normalized heating energy consumption = Actual heating energy consumption $\times$ CSI$_{normalized \, year}$/CSI$_{actual \, year}$.

Table 5.52 Mean ambient temperature

Year	Reference	2000
January	10.0	7.4
February	8.1	10.5
March	10.9	12.3
April	16.6	17.9
May	21.6	24.7
September	23.4	22.3
October	16.6	19.0
November	15.6	17.3
December	12.0	12.5

Table 5.53 Heating degree-days at 20°C for the heating season – normalization ratio $NRh = 1.146$

Year	Reference	2000
January	309	389
February	334	276
March	281	240
April	116	86
May	23	16
September	11	4
October	128	53
November	136	86
December	247	231
Season	1,584	1,382

Table 5.54 Cooling degree-days at 26°C for the cooling season – normalisation ratio $NRc = 0.634$

Year	May	June	July	August	September	Season
Reference	8	35	58	71	22	194
2000	0	56	131	98	20	306

Table 5.55 Values of the coefficient to use in the CSI relationship

	m_4	m_3	m_2	m_1	b
Winter	1.88	9.04E–06	−2.20	−0.002	0.96
Summer	−5.52	−6.62E–05	8.93	0.028	−4.81

Table 5.56 Cooling degree-days at 20°C – normalization ratio $NRc = 0.739$

Year	May	June	July	August	September	Season
Reference	73	157	202	230	113	775
2000	89	211	319	284	145	1,048

Table 5.57 CSI normalization ratio for heating – $NRh = 0.994$ and for cooling – $NRc = 0.980$

Year	DD_h	$(n/N)_h$	CSI_h	DD_c	$(n/N)_c$	CSI_c
Reference	296	0.507	0.506	176	0.792	1.698
2000	299	0.527	0.509	240	0.841	1.734

Table 5.58 Form table 1: Supplied energy

Supplied energy	Diesel oil	Electricity
Estimated (E) or measured (M)	M	M
Conversion factor	11.92 kWh/kg	1.00 kWh/kWh
November 1999–January 2000	7,212	
February–April 2000	8,666	
November 1999–February 2000		1,281
March–June 2000		1,292
July–October 2000		1,191
Sum (kWh/year)	15,877	3,764

Two electricity meters (one for the cheaper night rate) measure the whole consumption of the apartment. The bills based on these measurements are available every four months. Bills for the diesel oil are available at each delivery. The bills relating to the water consumption are available every three months.

Available data – calculation and results

The information given in Tables 5.58 and 5.60 is available from the bills, with the calculated system efficiencies given in Table 5.59, based on the hot-water supply temperatures given in Table 5.61.

Tables 5.62 and 5.63 give the actual and normalized (DD and CSI) values.

Remarks

Approximately 10 to 15% of the boiler energy production was used during the heating season for the production of hot water. During this period the heating was on for an average of nine hours per day.

From the above results we see that there is a considerable difference between the DD and CSI normalization results, especially during the cooling season. Since the CSI method takes into consideration both the average monthly degree-days and the solar availability, it is considered to be more suitable for countries like Greece, where weather data may vary substantially from year to year.

BUILDING NO. 2

General information

The building considered is a four-storey semi-detached building (Figure 5.21) constructed in 1992 and comprising eight apartments, with two on each floor. Only one apartment (situated on the third floor) is considered within this procedure.

The total heated area is equal to 91 m². Two adults and two children occupy the apartment. A natural ventilation system is used (opening windows).

A central heating is installed, running on diesel oil. The distribution of heat is realized via a water-borne system and radiators. The apartment is also equipped with a reversible heat pump used for auxiliary heating and cooling (split-unit air conditioner 2.8 kW rated power 1.4 kW – PER ≈ 2.0) running on electricity from 3 to 4 hours per day during hot summer days.

The hot-water production (approximately 45% of the total water consumption) occurs via a solar water heater with a back-up electrical heater for the winter season.

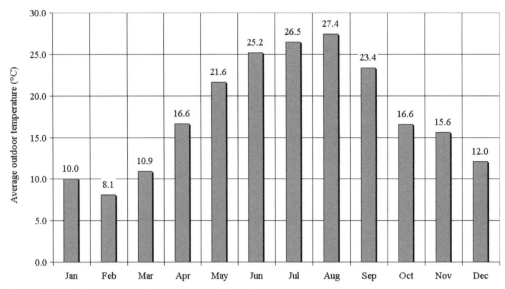

Figure 5.18 Monthly average outdoor temperature, Athens, Greece – normal year

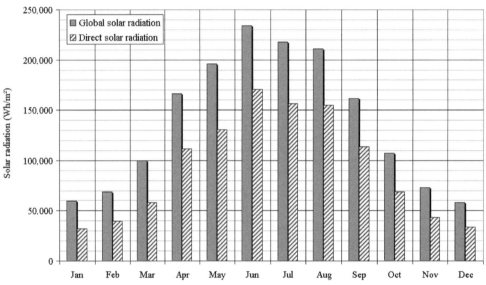

Figure 5.19 Monthly solar energy available on a horizontal surface, Athens, Greece – reference year

Figure 5.20 Greek building no. 1: Pictures of the building

Table 5.59 Form table 2: List of energy conversion systems, their efficiencies and how these were assessed

Energy conversion system	Fuel type	Winter efficiency	Summer efficiency	Estimated (*E*), measured (M) or calculated (C)
Household appliances and lighting	Electricity	1.00	1.00	*E*
Electrical boiler for hot water, 4 kW (1992)	Electricity		1.00	*E*
Central boiler for space and hot-water heating (1992)	Diesel oil	0.85		*E*
Split-unit heat pump for air conditioning	Electricity	1.00	COP 2.00	*E*

Table 5.60 Form table 3: Allocation of delivered energy, on an energy conversion system level, to each of the specific energies

Method	Footnote 1			Footnote 2	
Central heating system	Space heating	Space cooling	Appliances and lighting	Tap hot water	External appliances/spaces
October 1999–January 2000	5,429			701	
February–April 2000	6,626			740	
Sum	12,058			1,441	
Method		**Footnote 3**	**Footnote 4**	**Footnote 5**	
Electrical conversion systems	Space heating	Space cooling	Appliances and lighting	Tap hot water	External appliances/spaces
November 1999–February 2000			1,281		
March–June 2000		168	846	362	
July–October 2000		420	437	544	
Sum		588	2,564	906	
Total (kWh/year)	12,058	588	2,564	2,347	

1. Diesel oil quantities delivered for the central heating plant were 605 and 727 kg.

2. During the winter season the hot water is supplied by the central heating economizer unit (at around 60°C). i.e.

$$\sum W_c \, kg \times 4.18 \, kJ/kg\,K \times (60 - T_w) \, K/3{,}600 \, kJ/kWh$$

where $W_c = 4$ persons × 40 litres/day per person × number of days in the period and T_w is the mean water supply temperature in °C (see Table 5.61).

3. The energy for space cooling can only be estimated by the hours of use of the air-conditioning unit, i.e. $Q_{SpCelectric} = 2.8$ kW × 2 h/day × 30 days = 168 kWh, and $Q_{SpCelectric} = 2.8$ kW × 3 h/day × 50 days = 420 kWh (during this period there is a minimum of three weeks' vacation).

4. The energy for electrical appliances is estimated as the total delivered electricity minus the electricity for hot water and half of the electricity for space cooling.

5. During the summer, back-up electricity is used for hot water (minimum 50°C) i.e.

$$\sum W_c \, kg \times 4.18 \, kJ/kg\,K \times (50 - T_w) \, K/3{,}600 \, kJ/kWh$$

(During this period there is a minimum of three weeks vacation.)

Table 5.61 Calculation of the hot tap-water consumption

	Number of days	T_w (°C)	$Q_{hotwater}$ (kWh)
January	31	17	239
February	28	16	245
March	31	15	250
April	30	16	245
May	31	17	184
June	30	18	178
July	31	20	167
August	31	21	54
September	30	22	156
October	31	20	167
November	30	19	228
December	31	18	234

One electricity meter measures the whole consumption of the apartment. The bills, based on these measurements, are available every four months. Bills for the diesel oil are provided monthly by the building manager. The bills relating to the water consumption are available every 3 months.

Available data – calculation and results

The information given in Tables 5.64 and 5.66 is available from the bills, with the calculated system efficiencies given in Table 5.65, based on the hot-water supply temperatures given Table 5.67.

Tables 5.68 and 5.69 give the actual and normalized (DD and CSI) delivered and supplied energy consumptions.

Remarks

Approximately 65–70% of the annual hot-water energy demand is supplied by solar energy. This amount, which is included in Tables 5.68 and 5.69, is indirectly billed energy (actually prepaid), the cost of which is based on the investment capital and payback period.

During this heating period the system was on for an average of eight to nine hours per day.

From the above results we see that there is a considerable difference between the DD and CSI normalization results, especially during the cooling season. Since the CSI method

Table 5.62 Actual and normalized – delivered energy consumption

Year	Space heating kWh/year	Area-specific space heating kWh/m² · year	Space cooling kWh/year	Area-specific cooling kWh/m² · year	Global kWh/year	Area-specific global kWh/m² · year
2000	12,058	154.6	588	7.6	17,557	225.1
DD normal	13,818	177.2	435	5.6	19,164	245.7
CSI normal	11,986	153.7	576	7.4	17,473	224.0

Table 5.63 Actual and normalized – supplied energy consumption

Year	Space heating kWh/year	Area-specific space heating kWh/m² · year	Space cooling kWh/year	Area-specific cooling kWh/m² · year	Global kWh/year	Area-specific global kWh/m² · year
2000	14,181	181.8	294	3.8	19,641	251.8
DD normal	16,251	208.4	217	2.8	21,633	277.3
CSI normal	14,096	180.7	288	3.7	19,549	250.6

Figure 5.21 Greek building no. 2: Pictures of the building

Table 5.64 Form table 1: Supplied energy

Supplied energy	Diesel oil	Electricity
Estimated (E) or measured (M)	M	M
Conversion factor	11.92 kWh/kg	1.00 kWh/kWh
November 1999	2,385	
December 1999	5,278	
January 2000	2,640	
February 2000	5,191	
March 2000	2,253	
November 1999–February 2000		2,198
March–June 2000		2,079
July–October 2000		2,268
Sum (kWh/year)	17,748	6,545

takes into consideration both the average monthly degree-days and the solar availability, it is considered to be more suitable for countries like Greece, where weather data may vary substantially from year to year.

COMMENTS ON THE RESULTS FOR GREEK BUILDINGS 1 AND 2

The results from the two houses in the Athens area are comparable (when the similar construction, occupancy and weather conditions are considered). CSI normalized specific energy supplied for heating is around 186 kW/m² year, while the 36 kW/m² difference in the normal specific global

energy supplied to the detached house and the apartment (Table 5.70) can be attributed mainly to the inclusion of the solar energy, which was utilized for hot-water heating during the whole year.

A simple calculation of the heat-loss factors was also performed (Table 5.71), with the mean ambient temperature and a heating season of five months (3,600 hours) taken into account for the year 2000. The set-point temperature was used as the mean indoor temperature for each house. These were reported by the occupants; for the apartment this was set at 22°C because of the presence of an infant child.

The heat-loss factors turned out to be the same, although there is a 15% difference in the floor areas of the two houses and they are also exposed differently to the weather. A logical explanation could be that, as the apartment is twice as high above the ground as the detached house and is several kilometres away, it experiences different microclimatic conditions, which affect the energy performance of the building.

Conclusions concerning the BEP procedure

The BEP procedure has been applied to eight buildings in four countries. No major problems have been encountered when applying the protocol.

Because of the situations in the different countries, different normalization techniques have been applied within

Table 5.65 Form table 2: List of energy conversion systems, their efficiencies and how these were assessed

Energy conversion system	Fuel type	Winter efficiency	Summer efficiency	Estimated (E), measured (M) or calculated (C)
Household appliances and lighting	Electricity	1.00	1.00	E
Electrical/solar hot-water heater, 4 kW (1992)	Electricity	1.00		E
	Solar	0.40	1.00	E
Central boiler for space heating (1992)	Diesel oil	0.85		E
Split-unit heat pump for air conditioning	Electricity	1.00	COP 2.00	E

Table 5.66 Form table 3: Allocation of delivered energy, on an energy conversion system level, to each of the specific energies

Method	Footnote 1				
Central heating system	Space heating	Space cooling	Appliances and lighting	Tap hot water	External appliances/spaces
November 1999	2,027				
December 1999	4,487				
January 2000	2,244				
February 2000	4,412				
March 2000	1,915				
Sum	15,086				

Method		Footnote 2	Footnote 3	Footnote 4	
Electrical conversion systems	Space heating	Space cooling	Appliances and lighting	Tap hot water	External appliances/spaces
December 1999–March 2000			1,617	581	
April–July 2000		252	1,806	147	
August–November 2000		560	1,717	271	
Sum		812	5,141	998	

Method				Footnote 5	
Solar conversion systems	Space heating	Space cooling	Appliances and lighting	Tap hot water	External appliances/spaces
December 1999–March 2000				387	
April–July 2000				627	
August–November 2000				390	
Sum				1,405	
Total (kWh/year)	15,086	812	5,141	2,403	

1. Diesel oil quantities delivered for the central heating plant were 200.1, 442.8, 221.5, 435.5 and 189 kg.

2. The energy for space cooling can only be estimated by the hours of use of the air conditioning unit, i.e. $Q_{SpCelectric} = 2.8\,kW \times 3\,h/day \times 30\,days = 252\,kWh$, and $Q_{SpCelectric} = 2.8\,kW \times 4\,h/day \times 50\,days = 560\,kWh$ (during this period there is a minimum of three weeks vacation).

3. The energy for electrical appliances is estimated as the total delivered electricity minus the electricity for hot water and half of the electricity for space cooling.

4. During the winter the hot water is partly supplied by the solar heater – at least 40% of the load – with an electric heater as back-up (at around 60°C), i.e. $Q_{hotwater} \times 40\%$

$$Q_{hot\,water} = \sum W_c\,kg \times 4.18\,kJ/kg\,K \times (60 - T_w)\,K/3{,}600\,kJ/kWh$$

where $W_c = 4$ persons $\times$ 40 litres/day per person $\times$ number of days in the period and is the T_w mean water supply temperature in °C (see Table 5.67).

5. During the summer season 100% of the load is supplied by the solar heater (minimum temperature around 50°C), i.e.

$$Q_{hot\,water} = \sum W_c\,kg \times 4.18\,kJ/kg\,K \times (50 - T_w)\,K/3{,}600\,kJ/kWh$$

(During this period there is a minimum of three weeks vacation.)

the tests. Depending on the number of energy bills available during one year, the assessment of the heating/cooling energy consumption requires more or fewer assumptions. When only one bill per year is available, assumptions about the energy consumption for hot water or for cooking have to be made. If energy bills are available on a monthly basis, the analysis of the bills when neither heating nor cooling is applied makes it possible to identify the energy consumption for hot water or for cooking.

Estimations of the energy consumption have to be made when no bills are available, for instance when wood is burned.

In the BEP procedure, the real indoor climate is not taken into account and estimation of the system efficiency is performed.

The systematic application of the BEP procedure can easily be realized via the software that has been developed.

Table 5.67 Calculation of the hot tap-water consumption

	Number of days	T_W (°C)	$Q_{hotwater}$ (kWh)
January	31	17	239
February	28	16	245
March	31	15	250
April	30	16	245
May	31	17	184
June	30	18	178
July	31	20	167
August	31	21	54
September	30	22	156
October	31	20	167
November	30	19	228
December	31	18	234

Table 5.70 Energy supplied to the two Greek buildings

Building	$Q_{heating}$ kWh	q_H kWh/m²	Q_{global} kWh	q_G kWh/m²
Detached house	14,096	180.7	19,549	250.6
Apartment	17,642	193.9	26,163	287.5

Table 5.71 Heat-loss factors

Building	$Q_{delivered}$ kWh	T_s °C	T_a °C	HLF W/K
Detached house	11,986	20	12	416.2
Apartment	14,995	22	12	416.5

RESULTS FOR THE MONITORED ENERGY PROTOCOL (MEP) PROCEDURE

Introduction

The MEP protocol has been described extensively in Chapter 2 and is comparable with the BEP procedure. The following elements are different:

- Monitoring of the indoor and outdoor climate is carried out.
- Energy monitoring is carried out.
- The efficiency of the heating system is measured for most of the time.

Belgian buildings

BUILDING NO. 1

The monitoring carried out

Monitoring was carried out for 23 days in this house. The outdoor and indoor temperatures have been recorded. Energy monitoring based on manual observation has also been carried out during this period. Figure 5.22 presents the results of the temperature monitoring.

At the beginning of the monitoring, the occupants were not present.

The monitoring period was very cold; during the 23 days of the monitoring about 13% of the heating degree-days 15/15 of a normal heating season were registered. The

average indoor temperature during the monitoring period was 17.1°C. The average outdoor temperature was 2.5°C.

A summary of the results obtained during the energy monitoring is given in Table 5.72.

Available data – calculation and results

Based on the information from the monitoring, the new supplied energy consumption was calculated, as shown in Table 5.73.

A comparison between the results obtained in the BEP and MEP procedures is shown in Figure 5.23. The results are presented with an error band calculated by applying a variation of ±3°C to the correction factor for normalization.

As an extra service within MEP, a normalization of the indoor temperature is also carried out. Table 5.74 presents the results of the calculations performed with this new information:

1. Normalization with reference to the external climate takes into account the specific values monitored during the 23 days of the monitoring period. The normal DD have been calculated for this period.
2. Normalization with reference to the indoor climate assumes that the measured average indoor temperature (17.1°C) is equal to the average indoor temperature during the whole heating season. The degree-days calculated for this specific temperature are compared with the reference degree-days calculated for a house by applying a night-time setback (17°C average indoor temperature).

Table 5.68 Actual and normalized – delivered energy consumption

Year	Space heating kWh/year	Area-specific space heating kWh/m² · year	Space cooling kWh/year	Area-specific cooling kWh/m² · year	Global kWh/year	Area-specific global kWh/m² · year
2000	15,086	165.8	812	8.9	23,442	257.6
DD normal	17,289	190.0	600	6.6	25,433	279.5
CSI normal	14,995	164.8	796	8.7	23,335	256.4

Table 5.69 Actual and normalized – supplied energy consumption

Year	Space heating kWh/year	Area-specific space heating kWh/m² · year	Space cooling kWh/year	Area-specific cooling kWh/m² · year	Global kWh/year	Area-specific global kWh/m² · year
2000	17,748	195.0	406	4.5	26,277	288.7
DD normal	20,339	223.5	300	3.3	28,762	316.0
CSI normal	17,642	193.9	398	4.4	26,163	287.5

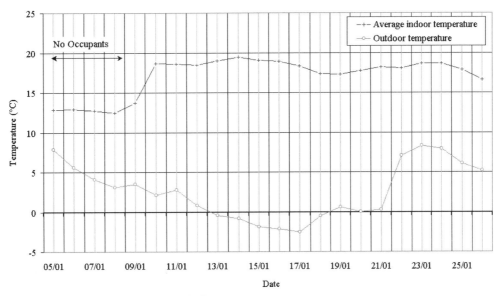

Figure 5.22 Results of the temperature monitoring

Table 5.72 Data obtained during the energy monitoring (in kWh)

		Energy conservation – Gas	Hot-water consumption	Cooking consumption	Q_{heat}	Energy consumption – electricity	Q_{tot}
Measured (M), estimated (E) or calculated (C)		M	E	E	C	M	C
Period	Number of days						
1	13.28	2,513.0	103.4	35.6	2,374.0	192.5	2,705.4
2	7.51	1,260.9	58.5	20.1	1,182.2	141.4	1,402.3

Table 5.73 Supplied normalized heating and global energy consumption (in kWh/year)

Year	$Q_{heat, sup, normal}$	$q_{heat, sup, normal}$	$Q_{tot, sup, normal}$	$q_{tot, sup, normal}$
Normal energy	27,585	101.0	35,796	131.1
	$27,172 < Q < 27,585$	$99.5 < Q < 101.0$	$35,384 < Q < 35,796$	$129.6 < Q < 131.1$

Table 5.74 Delivered normalized heating energy consumption

Year	$Q_{heat, del, normal}$ (kWh/year)	$q_{heat, del, normal}$ (kWh/m² year)
Normal external	16,551	60.6
Normal indoor	16,026	59

BUILDING NO. 2

The monitoring carried out

Monitoring was carried out in the house for 14 days:

- Manual energy monitoring was performed by the occupants and the different energy consumptions recorded.
- Monitoring of the indoor and outdoor climates was carried out simultaneously. Four temperatures inside the house were recorded as well as the outdoor temperature.

The results of the temperature monitoring are shown in Figure 5.24.

The average daily indoor temperature measured was 16.0°C. The average outdoor temperature during the monitoring was 0.3°C. The monitoring was carried out

during an extremely cold period for Belgium. The lowest average temperature on a daily basis was equal to −4.2°C. 10.6% of the normal degree-days have been registered during this 14-day period.

A summary of the results obtained during the energy monitoring is given in Table 5.75.

Available data – calculation and results

Based on the information from the monitoring, the new supplied energy consumption was calculated and is shown in Table 5.76.

A comparison between the results obtained from the BEP and MEP procedures is shown in Figure 5.25. The results are presented with an error band calculated by applying a variation of ±3°C to the correction factor for normalization.

As an extra service within MEP, a normalization of the indoor temperature is also carried out. Table 5.77 presents the results of the calculations performed with this new information:

1. The normalization with reference to the external climate takes into account the specific values monitored

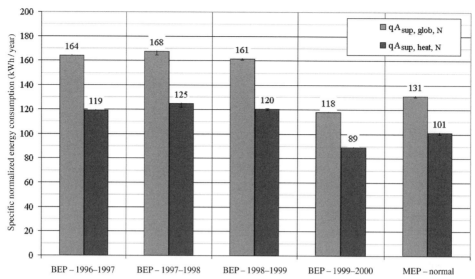

Figure 5.23 Comparison of the energy consumption

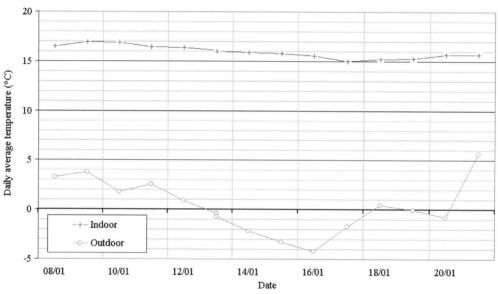

Figure 5.24 Results of the indoor climate monitoring

Table 5.75 Data obtained during the energy monitoring (in kWh)

		Energy consumption – gas	Hot-water consumption	Cooking consumption	Q_{heat}	Energy consumption – electricity	Q_{tot}
Measured (M), estimated (E) or calculated (C)		M	M	E	C	M	C
Period	Number of days						
1	6.55	643.7	50.9	25.7	567.1	63.7	712
2	7.14	767.4	51.4	28.1	688.0	125.1	903

Table 5.76 Supplied normalized heating and global energy consumption (in kWh/year)

Year	$Q_{heat, sup, normal}$	$q_{heat, sup, normal}$	$Q_{tot, sup, normal}$	$q_{tot, sup, normal}$
Normal external	11,795	42.7	21,020	76.1
	$11{,}795 < Q < 13{,}906$	$42.7 < Q < 50.4$	$21{,}020 < Q < 23{,}130$	$76.1 < Q < 83.6$

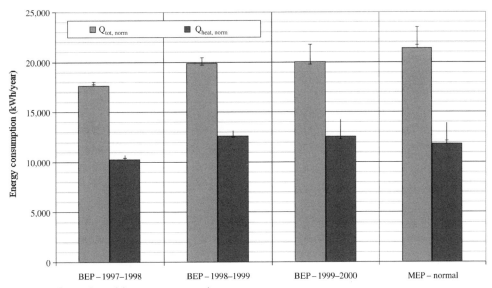

Figure 5.25 Comparison of the energy consumption

Table 5.77 Delivered normalized heating energy consumption

Year	$Q_{heat, del, normal}$ (kWh/year)	$q_{heat, del, normal}$ (kWh/m² year)
Normal external	15,765	57.1
Normal internal	18,817	68.1

during the 14-day monitoring period. The normal DD have been calculated for this period.

2. The normalization with reference to the indoor climate assumes that the measured average indoor temperature (16.0°C) is equal to the average indoor temperature during the whole heating season. The degree-days calculated for this specific temperature are compared with the reference degree-days calculated for a house by applying a night-time setback (17°C average indoor temperature).

Swedish buildings

BUILDING NO. 1

The monitoring carried out

Monitoring of the building was primarily concerned with measurement of the specific energies supplied by electricity and wood for the fan stove and with measurement of temperatures.

Electricity

The electricity suppliers make measurements on the overall electricity supplied. Hourly values are registered and these are collected on a daily basis by means of the telephone; at midnight the company automatically dials the customer number and data is silently transferred. These values can be seen on the bill.

With the overall electricity measured by the supplier, the task was to measure how this electricity was distributed within the residence. The strategy was to measure electricity that was supplied to the boiler units and, by subtracting these

values from overall values, assess what was supplied as specific energy for appliances and lighting.

The electric boiler is utilized for the purpose of space heating (10.5 kW) and for heating of tap water (9 kW). Measurements were conducted on the coils for tap-water heating, in other words, the energy delivered to the water was measured. Tap water is heated with three coils, each having a power of 3 kW. With the help of run-time sensors on each coil, the delivered energy could be assessed by multiplying by the power of each coil. As for the space-heating coils, measurement of the current supplied to the unit has been realized directly on the fuse.

Wood measurements

The amount of wood used for space-heating purposes was weighed and summed on a daily basis. The occupants performed the weighing.

Temperature measurements

Two thermal zones were predicted to exist on the basis that there were two storeys. The fan stove is situated on the lower storey next to an open staircase. Since the stove primarily blows warm air, it was considered that the upper storey would be warmer than the lower storey.

Measurements were conducted with thermistor sensors and a four-channel data logger. Two sensors were used for the thermal zones, a third for measurement of the external environment, while the fourth was placed in the air outlet of the fan stove.

Available data – calculation and results

Measurements were initiated on 17 January (day number 17) and terminated on 19 February (day number 50). A part of the series is shown in Figure 5.26. A temperature measurement was performed on the fan stove warm air outlet – a type of control for inspection of when it was in use. Contrary to prediction, the differences in air

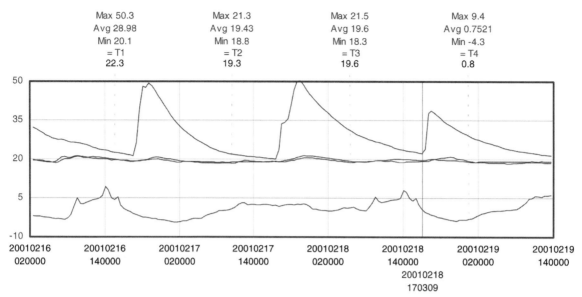

Figure 5.26 A sequence of four measured temperatures. T1 measures warm air from the fan stove. T2 is the temperature of the lower storey and T3 that of the upper storey. T4 is the external temperature

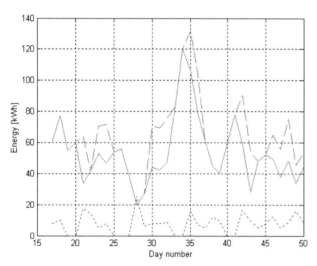

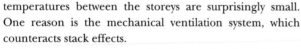

Figure 5.27 Monitored energy use (day number 17 = 17 January and day number 50 = 19 February). The dotted line represents daily energy for tap-water heating. The filled line corresponds to daily use of space heating and appliances. The dashed line is the sum of the latter energy use including the space heating delivered by the fan stove

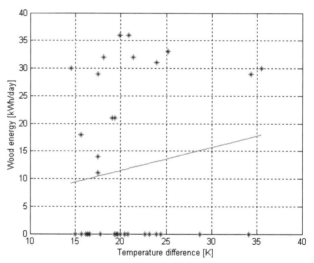

Figure 5.28 Wood use per day as a function of temperature difference. The spread of points does not show a clear correlation between use of wood and the temperature difference

temperatures between the storeys are surprisingly small. One reason is the mechanical ventilation system, which counteracts stack effects.

The occupants were away for four weekends, which can be seen in the tap-water heating in Figure 5.27. The rise afterwards is explained by laundry. The peak in space heating in the middle of the sequence is due to an extreme drop in temperature.

The recordings of the current to the space heater and to the tap-water heating unit were unfortunately lost. As a result of the intensive sampling frequency (1 second intervals), the batteries of the logger ran low and failed to maintain voltage sometime between the end of the monitoring and the attempted transfer of data to the computer.

Analysis of wood consumption

The occupants were requested to weigh the wood that was placed in the stove. Daily values were noted.

For wood consumption, two calculation schemes are shown. The first is made by assuming that the consumption of wood is climate dependent, and the second is based on the assumption that wood consumption is evenly distributed over the heating season.

As shown in Figure 5.28, the correlation between wood use and temperature difference is poor. The equation of the regression line is $(0.42 \times \text{Temp\_diff} + 3.1)$ kWh/day, which for the actual year gives a supplied wood heat corresponding to 2,843 kWh. However, this value includes wood burning during the summer, which is not practised. The value for the heating season will only be 2,427 kWh.

On the assumption that wood consumption is climate-independent, the daily mean corresponds to 11.9 kWh/day. With 243 days of heating (the default Swedish heating season length), the total is 2,880 kWh. This value may be somewhat high, when it is considered that the average daily value is derived from a period when the external temperature was on average −1.9°C.

Irrespective of which of the two methods was used, the results are in this case fairly consistent. The values estimated from the measurements do not seem to be unreasonable in comparison with the estimation by the occupant of the annual consumption (some 3,000 kWh).

Analysis of energy for tap-water heating

Tap-water heat was measured and recorded on a daily basis. With the assumption that this entity is constant throughout the heating season, a straightforward average can be calculated on the basis of measured data. The mean value is 7.6 kWh/day, giving an annual value of 2,782 kWh/year. This value may be overestimated for the following reasons:

- The incoming water temperature is higher during the summer period.
- Shower water temperatures are lower during the summer period.
- The residents are away for six weeks during the summer period.

The last point is to some extent taken into consideration in the average value, since the occupants were away during part of the monitoring period.

As mentioned in the discussion of BEP, the Swedish default value for one person is 1,000 kWh/person per year. The estimation above is higher for the two occupants. After an inquiry, it emerged that tap hot water is also supplied to the dishwasher, which leads to an increased value.

Estimates were made for the summer hot-water consumption. The period consists of 122 days, during which the occupants are away for six weeks. During the six weeks, the house is visited once a week. This gives 74 days at the house. Furthermore, a modification was made for the incoming water. Whereas the temperature is about 6°C during winter, the summer temperature is about 12°C. The set-point temperature is 80°C. Energy use for the summer period is estimated to be 523 kWh (74 days), and for the heating season 1,852 kWh (243 days). The annual energy for tap-water heating is estimated to be 2,375 kWh/year.

Tables 5.78 to 5.81 give details of the measured values, the system efficiencies and the non-normalized delivered energy.

In MEP, it is necessary that a calculation is done in which the indoor conditions are normalized, especially the indoor temperature. For this purpose, in accordance with using MUF as the normalization method, the standard set-point temperature was chosen to be 21°C (this type of value is

Table 5.78 Form table 1: Supplied energy

Supplied energy	Electricity	Wood
Estimated (E), measured (M) or calculated (C)	M	M, C
Conversion factor	1.00 kWh/kWh	1,500 kWh/m³
December 1999	1,874	
January 2000	1,847	
February 2000	1,459	
March 2000	1,426	
April 2000	1,040	
May 2000	688	
June 2000	509	
July 2000	474	
August 2000	438	
September 2000	640	
October 2000	769	
November 2000	963	
Heating season December 1999–November 2000	–	2,427
Sum (kWh/year)	12,127	2,427

prescribed on a national level). The values for delivered and supplied energy are given in Tables 5.82 and 5.83.

The heat-loss factor

The extended period may be used for estimation of the heat-loss factor. The 34 days of monitoring is short, when seen from the point of view of energy measurements. However, the period was relatively cold and large variations in the external temperature were recorded. The collected data was run through a static Energy Signature (ES) model. The effects of solar radiation and metabolic heat were neglected. The heat delivered by the radiators, appliances and lights, and combusted wood was regressed against temperature differences in the external and internal environments, on a daily basis. The results are plotted in Figure 5.29.

The points have a large scatter, mainly due to dynamic effects that would suggest the use of a dynamic model. The heat-loss factor, which corresponds to the slope of the regressed line, is found to be 140 W/K.

The next method to be applied is a static ES method that makes use of bills. With the assumption that the internal temperature is the same throughout the seasons as during the monitoring period, monthly energy use can be plotted against temperature difference. The results obtained are plotted in Figure 5.30.

The dashed line is regressed through the open circles. Here, global energy, excluding energy for hot water, is used. The heat-loss factor is 113 W/K on the basis of all the points. Note the shift in the four left-most points, which depict energy use during the summer: these can be excluded from the regression. Exclusion of these gives a heat-loss factor of 132 W/K. If only the last three points are omitted, the value becomes 124 W/K.

The line with open diamonds is a regression with specific energy for space heating, as shown in the previous tables. These points indicate a heat-loss factor corresponding to 134 W/K.

Table 5.79 Form table 2: List of energy conversion systems, their efficiencies and how these were assessed

Energy conversion system	Fuel type	Winter efficiency	Summer efficiency	Estimated (E), measured (M) or calculated (C)
Household appliances and lighting	Electricity	1.00	1.00	E, C
Electrical boiler for space heating and hot water, 10.5 kW (1982)	Electricity	1.00	1.00	M, C
Fan stove, wood manual feeding (installed 1993)	Wood	0.85	0.85	E (manufacturer info)

Table 5.80 Form table 3: Allocation of delivered energy, on an energy conversion system level, to each of the specific energies

Method	Footnote 3		Footnote 2	Footnote 1	
Electrical conversion systems	Space heating	Space cooling	Appliances and lighting	Tap hot water	External appliances/spaces
December 1999	1,243		394	236	0
January 2000	1,216		394	236	0
February 2000	900		356	213	0
March 2000	796		394	236	0
April 2000	430		382	229	0
May 2000	110		394	184	0
June 2000	0		382	128	0
July 2000	0		342	132	0
August 2000	0		306	132	0
September 2000	80		382	178	0
October 2000	139		394	236	0
November 2000	353		382	229	0
Sum	5,256		4,502	2,369	0
Method	**Footnote 4**				
Wood fan stove	Space heating	Space cooling	Appliances and lighting	Tap hot water	External appliances/spaces
December 1999	309				
January 2000	304				
February 2000	264				
March 2000	273				
April 2000	210				
May 2000	79				
June 2000	0				
July 2000	0				
August 2000	0				
September 2000	81				
October 2000	175				
November 2000	209				
Sum	1,904				
Total (kWh/year)	7,160	0	4,502	2,369	0

1. Electricity used to heat the tap water was measured. The daily mean value for the heating season was determined to be 7.6 kWh/day. For the summer period (15 May to 15 September), the mean value was modified, since the occupants were present some 78 days out of the 122 possible, and a correction term, (80 − 12)/(80 − 6), was applied to allow for incoming water temperature rising from 6 to 12°C.

2. The data for assessing the energy used by appliances was lost. Relying on bills, the month of June was used to assess the mean appliance power by using the electricity bill value (global) and subtracting the energy for heating water. The difference was divided by month hours, yielding 530 W. For July and August, the appliance energy is the billed value reduced by the hot-water energy.

3. The space-heating energy delivered from the radiator system is estimated to be the total delivered electricity minus the electricity for tap-water heating (footnote 1) and the electricity for the appliances (footnote 2).

4. From measurements, the wood energy content was regressed against the temperature difference between the internal and external environments. The regression line was used, with mean monthly temperatures, to estimate monthly consumption. For the months May and September, the value was halved because of the definition of the heating season. The total amount of wood was estimated by the occupant to be 2 m³. MEP measurements and calculations estimated 1.6 m³.

The solid line with regression on the stars is based on the heat losses at the set-point temperature (the EN 832 type of calculation). In other words, only the useful heat gains and space-heating energy are plotted for the entire year. The heat-loss factor with this method is 140 W/K. The fact that this value is exactly the same as from the monitored period is probably coincidental.

From this analysis, the heat-loss factor lies in the range between 130 and 140 W/K.

Table 5.81 Form table 4: Display of chosen processing periods, actual delivered space-heating energy for the periods, numerical values of influencing variables, and normalized space-heating energy. The lowest two rows indicate the band (minimum and maximum) values that are obtained with prescribed conditions from the normalization methods. The method used for normalisation was MUF

Period	Space heating	Actual exterior temp	Actual direct solar	Actual diffuse solar	Normal exterior temp	Normal direct solar	Normal diffuse solar	Normal space heating
December 1999	1,552	−1.6	16,484	4,860	−1.9	13,600	4,500	1,585
January 2000	1,521	−1.2	33,285	7,661	−3.7	18,300	7,400	1,786
February 2000	1,534	−0.1	48,482	16,842	−3.8	38,000	16,400	1,479
March 2000	1,069	1.6	138,989	31,982	−0.5	68,400	34,400	1,320
April 2000	640	6.7	96,270	51,972	4.2	103,700	54,500	863
May 2000	188	12.2	150,266	75,099	10.4	169,000	68,500	320
June 2000	0	14.3	137,121	87,221	15.1	146,800	76,000	5
July 2000	0	16.3	102,765	69,320	16.7	156,700	72,500	4
August 2000	0	15.9	131,930*	34,670*	15.7	121,800	61,400	8
September 2000	162	11.1	150,235	36,016	11.3	92,700	37,400	169
October 2000	313	10.5	18,314	20,890	7.0	49,800	21,300	571
November 2000	562	6.8	3,286	5,965	1.8	21,300	8,400	938
Sum	7,160							9,050
							Minimum	8,919
							Maximum	9,827

*Not available; reference year 1971 data used for August.

Table 5.82 Delivered energy. Normalized values were obtained with MUF

Year	Space heating kWh/year	Area-specific space heating kWh/m² · year	Space cooling kWh/year	Area-specific space cooling kWh/m² · year	Global kWh/year	Area-specific global kWh/m² · year
December 1999–November 2000	7,159	60.2	0	0	14,030	117.9
Normal external	9,050	76.0	0	0	15,921	133.8
Normal internal	10,696	89.9	0	0	17,567	147.6

Table 5.83 Supplied energy. Normalized values were obtained with MUF

Year	Space heating kWh/year	Area-specific space heating kWh/m² · year	Space cooling kWh/year	Area-specific space cooling kWh/m² · year	Global kWh/year	Area-specific global kWh/m² · year
December 1999–November 2000	7,534	63.3	0	0	14,366	120.7
Normal external	9,475	79.6	0	0	16,346	137.4
Normal internal	11,197	94.1	0	0	18,068	151.8

Remarks

The MEP results differ from those from BEP. The largest effect is due to estimations in the non-billed wood consumption. In BEP, where the occupant supplied the information, wood consumption seems to be overestimated. On the other hand, the estimates in MEP may be erroneous since there is no clear information on whether the use of wood is dependent on or independent of climate. What MEP does is estimate the order of magnitude, providing a type of verification as to whether or not the information supplied is correct.

Though the extended period can be used to determine the heat-loss factor of the building, there may be circumstances where the frequency and quality of the bills are such that, in combination with short-term MEP, they allow evaluation of this value. This has been shown for both Swedish buildings. It has been possible because the billing frequency is monthly and temperatures have been assessed.

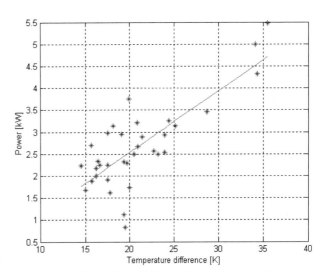

Figure 5.29 Static model applied to daily mean values. The power delivered by heating units and appliances is plotted against temperature difference. The effect of solar radiation is neglected

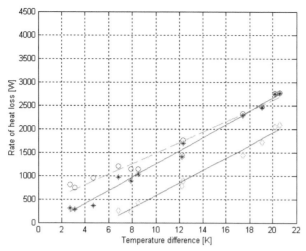

Figure 5.30 Regression lines for energy use against temperature difference. The diamonds are data for delivered specific space heating. The open circles represent global delivered energy excluding hot-water energy. Stars depict calculated heat losses at the set-point temperature

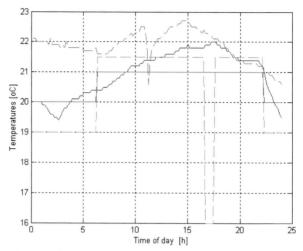

Figure 5.31 The dashed lines show the set-point temperature and room temperature for the thermostat in the south-facing room (see plan drawing in Figure 5.32). The filled lines are for corresponding temperatures in a north-facing room

BUILDING NO. 2

This building is fully equipped with monitoring equipment for the purpose of measuring specific energy. The system can measure the temperature at the location of the thermostat controls, but this facility had not previously been applied. Sampling was carried out every 2 seconds.

Specific energy is continuously measured and billed. The only information that the standard procedures of MEP will add to that provided by BEP relates to temperature. This means that processed values of BEP do not change. The reasons are as follows:

- Actual specific energy use does not change; the values of this have been measured and billed for over a year.
- Normalization calculations with MUF make use of set-point temperatures, not actual temperatures.
- The heat-loss factor is not affected by a systematic change in internal temperature.

The monitoring carried out

The apartment is equipped with two thermostat controls, one in the southern thermal zone and the other in the northern zone. These were now triggered to register values. Intensive measurements were performed for a period of two weeks, with values saved every 11.25 minutes. The temperatures for 3 April 2001 are shown in Figure 5.31.

The purpose of the measurements is to verify that the set-point temperatures are correct. Note that the daytime set-point is higher than in the BEP analysis. The reason for this is that, during the audit, the occupants noted that the apartment was chilly during cold winter periods. The auditor showed the occupants how the set-point temperature could be reset for daytime. The set-point temperature was raised by 1°C to 21.5°C. Since the MEP measurements, the occupants have not changed the set-point temperatures again.

Available data – calculation and results

The results of measurement and calculation are given in Tables 5.84 to 5.87, with the actual and normalized values of the energy in Tables 5.88 and 5.89. The energy values are not affected by temperature.

Remarks

For all intents and purposes, the quality of billed data for a building with this type of billing/monitoring system is such that the standard procedures of MEP (which exclude the extra services) are superfluous. The extra information provided in this case was the internal temperatures. The system was capable of recording this data, but had not been used in this way prior to the MEP measurements.

The results will therefore be the same as for the BEP procedure. The heat-loss factor, based on measured (billed) monthly values, will not be influenced by the mean temperature since the slope of the line (delivered heat plotted against temperature difference) is unaffected.

In general, the energy analysis for individual modern Swedish multi-family apartments may tend to become complicated. With envelope elements that are well insulated, transmission may be quite small, which leads to a larger influence of heat exchange between apartments, via both transmission and air leakage. The temperatures of neighbouring apartments and the characteristics of the internal construction may have to be taken into consideration in the analysis in order to obtain 'actual' values of energy use, especially for space heating.

Spanish buildings

BUILDING NO. 1
The monitoring carried out

A period of 40 days, from 27 January to 8 March, was selected as representative of the monitoring carried out in this house. During this period the outdoor and indoor temperatures, the

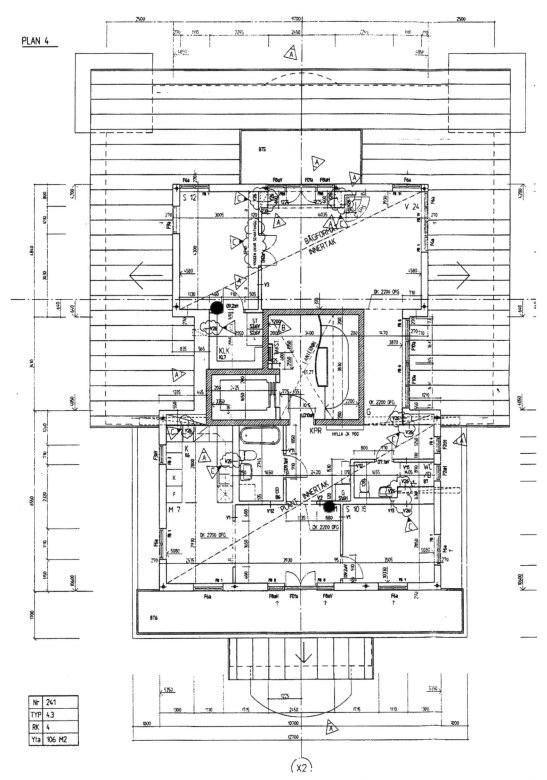

Figure 5.32 Plan drawing of the apartment; the downwards direction is south. The thermostat controls, where set-point temperatures are actuated and room temperatures collected, are indicated by dots. The zone within the shaded boundary is the corridor and elevator, i.e. common premises

solar radiation and the electricity consumption have been registered. The electricity consumption was base on manual observation.

Figure 5.33 shows the average indoor and outdoor temperatures in this period. From this figure it can be seen how short periods of three days have been grouped together in order to obtain results and to correlate

energy consumption with the indoor–outdoor temperature difference and the solar radiation.

Available data – calculation and results

A summary of the results obtained during the energy monitoring is given in Table 5.90.

Table 5.84 Form table 1: Supplied energy

Supplied energy	Electricity (appliances)	District heating 1	District heating 2	District + electricity
Estimated (E) or measured (M)	M	M	M	M
Conversion factor	1.00 kWh/kWh	1.00 kWh/kWh	1.00 kWh/kWh	1.00 kWh/kWh
January 2000	573	2,497	197	656
February 2000	503	2,135	169	589
March 2000	482	1,830	186	582
April 2000	410	949	137	463
May 2000	406	229	202	152
June 2000	369	66	175	230
July 2000	302	0	164	111
August 2000	300	14	164	171
September 2000	329	158	148	208
October 2000	423	575	180	319
November 2000	399	1,296	142	356
December 2000	490	1,760	208	447
Sum (kWh/year)	4,986	11,509	2,072	4,284

Table 5.85 Form table 2: List of energy conversion systems, their efficiencies and how these were assessed

Energy conversion system	Fuel type	Winter efficiency	Summer efficiency	Estimated (E), measured (M) or calculated (C)
Household appliances and lighting	Electricity (appliances)	1.00	1.00	E
District heating 1 + 2; 1 = space heating, 2 = hot water	District heating	1.00	1.00	E
District heating and electricity of external spaces and appliances	District heating + electricity	1.00	1.00	E

These values have been used to obtain the heat-loss factor UA in W/K and the solar aperture gA value in m^2, through a linear correlation for which $R^2 = 0.81$. Therefore, the energy consumption can be estimated using the following correlation:

$$\text{Heating energy consumption} = UA(T_{in} - T_{out}) + gA \text{ Rad}$$

where for this building $UA = 227\,\text{W/K}$, $gA = -1.45\,\text{m}^{2*}$.

Figure 5.34 shows the correlation between the results and the measured data.

The delivered energy for heating can be estimated by taking into account that the temperature difference $(T_{in} - T_{out})$ is equal to the heating degree-days for the winter season. The number of heating degree-days for this period is 417. At the same time, the total radiation over a horizontal surface is 395 kWh/m^2 if the total radiation for those days with an average temperature less than 15°C is added. Thus, the heating energy consumption is 1,698 kWh (18.87 kWh/m^2).

Greek buildings

INTRODUCTION

During the two-week monitoring period the fuel and energy consumptions were recorded, together with the indoor and outdoor temperatures and the indoor humidity, as well as the occupants' habits.

Physically, $g.A$ value should always be greater than 0.

*The negative $g.A$ value is due to the mathematical analysis applied on the input data where an inaccuracy exists.

BUILDING NO. 1

The monitoring carried out

The indoor temperature and humidity sensors were placed in the living room and in one of the bedrooms (see Figure 5.35). The outdoor temperature sensor was located at a sheltered position on the balcony. The sensors (Tinytags) were installed on 31 January and were removed on 28 February 2001; data were recorded every 30 minutes. The detailed monitoring of energy use in the house was carried out from 10 to 23 February 2001. The temperature profiles are shown in Figure 5.36. During this period the average ambient temperature was 11.3°C and the mean indoor temperature 19.9°C.

Available data – calculation and results

From the monitoring data the uncertainty in the set-point control and in the indoor air temperature measurements may vary between ±1 and 3°C. For this reason normalization ratios were estimated for an indoor temperature of 20°C with an error band of ±2°C. Degree-days were calculated from the February measurements. Values are given in Tables 5.91 to 5.94.

Normalization with respect to indoor climate was done for an air temperature of 19.9°C – the average during the monitoring period.

Remarks

Looking at the profile of the indoor air temperatures, we see that this follows the pattern of the ambient air, with a relatively high amplitude. This is because of the low

Table 5.86 Form table 3: Allocation of delivered energy, on an energy conversion system level, to each of the specific energies

Method			Footnote 1		
Household appliances and lighting	Space heating	Space cooling	Appliances and lighting	Tap hot water	External appliances/spaces
January 2000			573		
February 2000			503		
March 2000			482		
April 2000			410		
May 2000			406		
June 2000			369		
July 2000			302		
August 2000			300		
September 2000			329		
October 2000			423		
November 2000			399		
December 2000			490		
Sum			4,986		

Method	Footnote 2			Footnote 3	
District heating 1 + 2	Space heating	Space cooling	Appliances and lighting	Tap hot water	External appliances/spaces
January 2000	2,497			197	
February 2000	2,135			169	
March 2000	1,830			186	
April 2000	949			137	
May 2000	229			202	
June 2000	66			175	
July 2000	0			164	
August 2000	14			164	
September 2000	158			148	
October 2000	575			180	
November 2000	1,296			142	
December 2000	1,760			208	
Sum	11,509			2,072	

Method					Footnote 4
District heating + Electricity	Space heating	Space cooling	Appliances and lighting	Tap hot water	External appliances/spaces
January 2000					656
February 2000					589
March 2000					582
April 2000					463
May 2000					152
June 2000					230
July 2000					111
August 2000					171
September 2000					208
October 2000					319
November 2000					356
December 2000					447
Sum					4,284
Total (kWh/year)	11,509	0	4,986	2,072	4,284

1. Electricity delivered to the apartment is measured and billed at the end of each month.

2. Delivered space heating for the apartment is measured and billed at the end of each month.

3. The quantity of tap water, hot and cold separately, is measured. The system operator calculates a temperature increase for hot water corresponding to 55°C times the specific heat capacity of water. This quantity is billed.

4. External energy use (measured) is for common-space heating, lighting, laundries and operation of installations (fans, pumps and elevators).

Table 5.87 Form table 4: Display of chosen processing periods, actual space-heating energy for the periods, numerical values of influencing variables, and normalized space-heating energy. The bottom two rows indicate the band (minimum and maximum) values that are obtained with conditions prescribed according to the MUF normalization method

Period	Space heating	Actual external temperature	Actual direct solar	Actual diffuse solar	Normal external temp	Normal direct solar	Normal diffuse solar	Normal space heating
January 2000	2,497	−1.2	33,285	7,661	−3.7	18,300	7,400	2,940
February 2000	2,135	−0.1	48,482	16,842	−3.8	38,000	16,400	2,722
March 2000	1,830	1.6	138,989	31,982	−0.5	68,400	34,400	2,299
April 2000	949	6.7	96,270	51,972	4.2	103,700	54,500	1,206
May 2000	229	12.2	150,266	75,099	10.4	169,000	68,500	325
June 2000	66	14.3	137,121	87,221	15.1	146,800	76,000	361
July 2000	0	16.3	102,765	69,320	16.7	156,700	72,500	41
August 2000	14	15.9	131,930*	34,670*	15.7	121,800	61,400	140
September 2000	158	11.1	150,235	36,016	11.3	92,700	37,400	183
October 2000	575	10.5	18,314	20,890	7.0	49,800	21,300	926
November 2000	1,296	6.8	3,286	5,965	1.8	21,300	8,400	1,931
December 2000	1,760	−1.6	16,484	4,860	−1.9	13,600	4,500	2,346
Sum	11,509							14,932
							Minimum	14,200
							Maximum	16,067

Table 5.88 Delivered energy. Normalized values were obtained with MUF

Year	Space heating kWh/year	Area-specific space heating kWh/m² · year	Space cooling kWh/year	Area-specific space cooling kWh/m² · year	Global kWh/year	Area-specific global kWh/m² · year
2000	11,509	108.6	0	0	18,567	175.2
Normal	14,932	140.9	0	0	21,990	207.5

Table 5.89 Supplied energy. Normalized values were obtained with MUF

Year	Space heating kWh/year	Area-specific space heating kWh/m² · year	Space cooling kWh/year	Area-specific space cooling kWh/m² · year	Global kWh/year	Area-specific global kWh/m² · year
2000	11,509	108.6	0	0	22,851	215.6
Normal	14,932	140.9	0	0	26,274	247.9

thermal mass of the building and the intermittent heating and because the daily morning ventilation is accomplished by opening the windows. During the heating operation the room temperature was always above the set-point temperature (21 to 23°C).

Approximately 13% of the boiler energy production was used for the production of hot water during this two-week period. During this period the heating was on for 126 hours (9 hours per day).

Since the CSI method takes into consideration both the average monthly degree-days and the solar availability, it was considered to be more suitable for normalization in Greece.

BUILDING NO. 2

The monitoring carried out

The indoor temperature and humidity sensors were placed in the living room and in one of the bedrooms (see Figure 5.37). The outdoor temperature sensor was located at a sheltered position on the balcony. The sensors (Tinytags) were installed on 27 January and were removed on 2 March 2001; data were recorded every 30 minutes. The detailed monitoring of energy use in the house was carried out from 15 to 28 February 2001. During this period the average ambient temperature was 12.3°C and the mean indoor temperature 20.6°C. The temperature profiles are shown in Figure 5.38.

Available data – calculation and results

From the monitoring data the normalization ratios were estimated for an indoor temperature of 20°C with an error band of ±2°C. Degree-days were calculated from the February measurements. Values are given in Tables 5.95 to 5.98.

From the monitoring data the uncertainty in the set-point control and in the indoor air temperature measurements may vary between ±1 and 3°C. For this reason normalization ratios were estimated for an indoor temperature of 20°C with an error band of ±2°C.

Normalization with respect to indoor climate was done for an air temperature of 20.6°C – the average during the monitoring period.

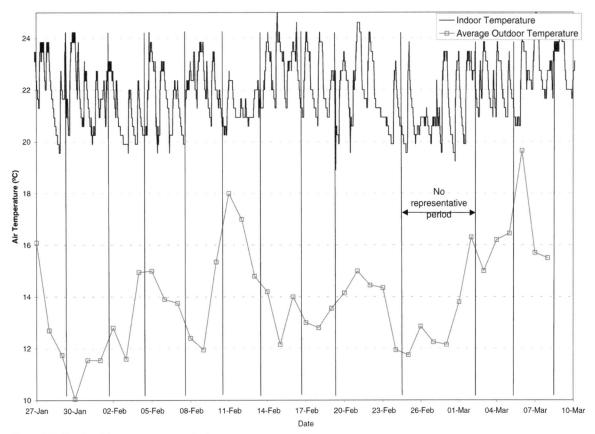

Figure 5.33 Results of the temperature monitoring

Table 5.90 Data obtained during the energy monitoring

		Energy consumption (W)	Energy consumption – appliances (W)	Q_{heat} (W)	$T_{in} - T_{out}$ (K)	Solar radiation (W)
Measured (M), estimated (E) or calculated (C)		M	E	C	M	M
Period	Number of days					
1	3	1,936	194	1,742	8.62	111
2	3	2,394	194	2,200	10.78	140
3	3	1,692	194	1,497	7.87	133
4	3	1,631	194	1,436	7.37	66
5	3	2,349	194	2,154	8.83	133
6	3	974	194	779	4.94	179
7	3	2,074	194	1,879	9.37	146
8	3	1,936	194	1,742	9.09	166
9	5	1,597	194	1,403	7.75	135
10	3	1,524	194	1,329	6.11	71
11	3	1,753	194	1,558	6.13	116

Remarks

Looking at the profile of the indoor air temperatures, we see that this follows the pattern of the ambient air, with a relatively high amplitude. This is because of the low thermal mass of the building and the intermittent heating and because the daily morning ventilation was accomplished by opening the windows. During the heating operation the room temperature was always above the set-point temperature (22 to 24°C).

Approximately 40% of the hot-water energy demand was supplied by solar energy during this two-week period.

During this period the heating was on for 83 hours (about 6 hours per day).

Since the CSI method takes into consideration both the average monthly degree-days and the solar availability, it was considered to be more suitable for normalization in Greece.

COMMENTS ON BEP AND MEP RESULTS FOR THE TWO GREEK BUILDINGS

The MEP results from the two houses in the Athens area are comparable to the BEP results. For the detached house

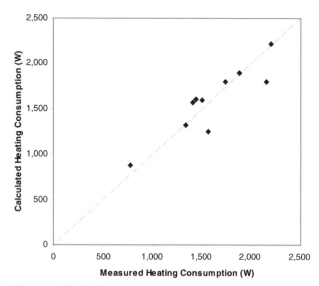

Figure 5.34 Correlation of the results with the measured data

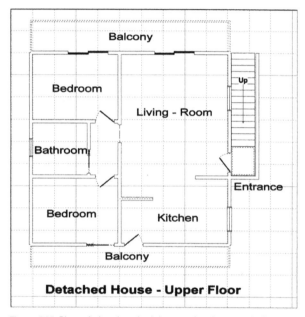

Figure 5.35 Plan of the detached house; the downward direction is south-west

Table 5.92 Normalization ratios for heating – NR @ 22 = 1.610, NR @ 20 = 1.575 and NR @ 18 = 1.466

Period	DD @ 22	DD @ 20	DD @ 18	(n/N)	CSI @ 20
Reference year	277	297	390	0.507	0.505
February 2001	172	232	266	0.583	0.293

daily in 2000) because of the presence of a new-born baby (while during the monitoring period in February 2001 this was reduced by 30% to 6 hours daily), thus leading to a lower fuel consumption and energy results. The results are summarized in Table 5.99.

A simple calculation of the heat-loss factors was also performed with the mean ambient temperature and the heating delivered during the 14-day period (336 hours) taken into account (Table 5.100). The average recorded temperature was used for the mean indoor temperature of each house, although the set-point temperature was 20°C in both cases.

From the average outdoor temperatures, the effect of microclimate is obvious, since the two houses in Athens, which are only a few kilometres apart, are exposed to different weather conditions. The heat-loss factor values found above are close to the 416 W/K (from BEP) and the difference can be attributed to the fact that the detached house is more exposed being the top floor (up a hill with a lot of trees all around), when compared to the apartment, which is on an intermediate floor (in a heavily built area), although the floor area of the latter is 15% higher.

Conclusions concerning the MEP procedure – comparison with BEP

The major difference in the MEP procedure when compared to the BEP procedure is that the accuracy of the final results must be better.

The application of the MEP procedure requires an energy monitoring to be performed. Several type of energy monitoring have been applied:

- monitoring of the temperature (indoor and outdoor temperature)
- energy monitoring, carried out on the basis of automatic recording of the consumption or on the basis of manual readings of the gauges.

The MEP procedure delivers results for the normalized energy consumption based on the outside climate, as well

the CSI normalized specific energy supplied for heating is around 190 kWh/m² year, while the normalized global energy supplied is 265 kWh/m² year.

The difference between the MEP (2001) and BEP (2000) results for the apartment house, can be attributed mainly to the higher use of the heating system (around 9 hours

Table 5.91 Data obtained during the energy monitoring (in kWh)

		Energy consumption – oil	Hot water consumption	Cooking consumption	Q_{heat}	Energy consumption – electricity	Q_{tot}
Measured (M), estimated (E) or calculated (C)		M	E	E	C	M	C
Period	Number of days						
1	7	981.6	67.7	24.6	921.9	26.9	1,041.1
2	7	754.7	72.1	20.5	688.6	56.0	837.2
Total	14	1,736.3	139.8	45.1	1,610.5	82.9	1,878.3

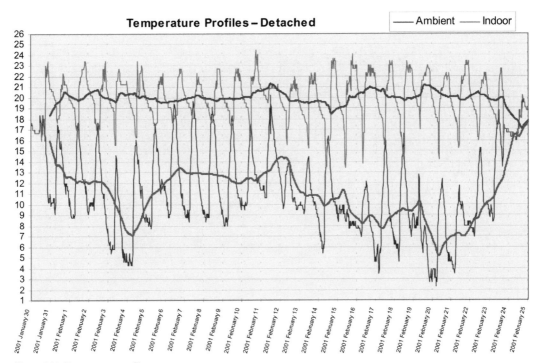

Figure 5.36 Temperature profiles

Figure 5.37 Plan of the apartment; the downward direction is south-east

Table 5.93 Supplied normalized heating and global energy consumption (in kW/year)

Year	$Q_{heat,\,sup,\,normal}$	$q_{heat,\,sup,\,normal}$	$Q_{tot,\,sup,\,normal}$	$q_{tot,\,sup,\,normal}$
Normal external	16,146	206.9	21,753	278.9
	$15{,}028 < Q < 16{,}505$	$192.7 < q < 211.6$	$20{,}635 < Q < 22{,}112$	$264.6 < q < 283.5$

Table 5.94 Delivered normalized heating energy consumption (in kW/year)

Year	$Q_{heat,\,del,\,normal}$	$q_{heat,\,del,\,normal}$
Normal external	13,724	175.9
Normal internal	13,655	175.1

Table 5.95 Data obtained during the energy monitoring (in kWh)

	Energy consumption – oil	Hot water consumption	Cooking consumption	Q_{heat}	Energy consumption – electricity	Q_{tot}	
Measured (M), estimated (E) or calculated (C)	M	E	E	C	M	C	
Period	Number of days						
1	7	726.3	114	32	726.3	27.5	899.8
2	7	628.1	133	52	628.1	44.5	857.6
Total	14	1,354.4	247	84	1,354.4	72.0	1,757.4

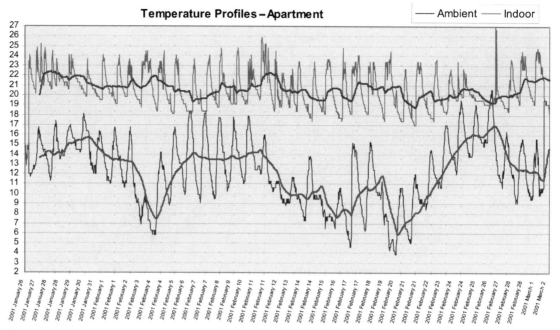

Figure 5.38 Temperature profiles

Table 5.96 Normalization ratios for heating – NR @ 22 = 1.583, NR @ 20 = 1.575 and NR @ 18 = 1.359

Period	DD @ 22	DD @ 20	DD @ 18	(n/N)	CSI @ 20
Reference year	277	297	390	0.507	0.505
February 2001	175	232	287	0.583	0.293

Table 5.97 Supplied normalized heating and global energy consumption (in kW/year)

Year	$Q_{heat, sup, normal}$	$q_{heat, sup, normal}$	$Q_{tot, sup, normal}$	$q_{tot, sup, normal}$
Normal external	15,111	166.1	23,574	259.1
	$13,039 < Q < 15,284$	$143.3 < q < 167.9$	$21,502 < Q < 24,737$	$236.3 < q < 260.9$

Table 5.98 Delivered normalized heating energy consumption (in kW/year)

Year	$Q_{heat, del, normal}$	$q_{heat, del, normal}$
Normal external	12,844	141.1
Normal internal	13,229	145.3

Table 5.99 Comparison of the results of MEP and BEP methods

Building	Method	$Q_{Heating}$ (kWh)	$q_{Heating}$ (kWh/m²)	Q_{Global} (kWh)	q_{Global} (kWh/m²)
Detached	MEP	16,146	206.9	21,753	278.9
house	BEP	14,096	180.7	19,549	250.6
Apartment	MEP	15,111	166.1	23,574	259.1
	BEP	17,642	193.9	26,163	287.5

Table 5.100 Heat-loss factors

Building	$Q_{Delivered}$ (kWh)	T_s (°C)	T_a (°C)	HLF (W/K)
Detached house	1,284	19.9	11.3	444
Apartment	1,151	20.6	12.3	413

as a normalized energy consumption based on the indoor climate.

The different normalization techniques available have been applied in these tests (degree-day, Climate Severity Index, modified utilization factor).

AUTHORS

The following authors have participated in the preparation of this chapter. The monitoring and calculations have been realized at a national level.

Belgium

Xavier Loncour, Belgian Building Research Institute (BBRI – CSTC – WTCB), Boulevard Poincaré 79, B-1060-Brussels, Belgium; e-mail: xavier.loncour@bbri.be

Greece

George Zannis and Mat Santamouris, University of Athens, Physics Department, Building & Environmental Studies Group, Panepistimioupolis, Athens 157 84, Greece; e-mail: gzannis@phys.uoa.gr, msantam@cc.uoa.gr

Spain

Servando Alvarez, AICIA Universidad de Sevilla, Departamento de Ingeniería Energética y Mecánica de Fluidos, Avenida Reina Mercedes, s/n-Apdo n° 1182, E-41080 Seville, Spain; e-mail: SAD@tmt.us.es

Sweden

Jan Akander, Kungliga Tekniska Högskolan (KTH), Department of Building Sciences, Division of Building Technology, Brinellv 34, SE-100 44 Stockholm, Sweden; e-mail: akander@bim.kth.se

REFERENCES

1. NIT 212, 1999, *Code de bonne pratique pour l'installation des chauffes-eaux solaires*. Belgian Building Research Institute, Brussels.
2. BBRI, 1999, Internal document – EPIGOON project – werkdocument 2.007. *Ontwerpnota betreffende EPN, algemene benadering*. Belgian Building Research Institute, Brussels.
3. Persson T, 2000, *Measurements of solar radiation in Sweden 1983–1998*. Reports Meteorology and Climatology. Swedish Meteorological and Hydrological Institute.

APPENDIX 1

Audit form

ADMINISTRATIVE INFORMATION

Municipality   House number   ID number

House address: _____

Estate property ID   Coordinates X, Y   Number of buildings

Owner/administrator: _____

Owner address (if other than house address): _____

CONTACT PERSON

Building   Name: _____

                Address: _____ Tel: _____

                Available to be contacted (days and time): _____

                Agreed time for audit (date and time): _____

Apartment   Name: _____

                Address: _____ Tel: _____

                Available to be contacted (days and time): _____

                Agreed time for audit (date and time): _____

Audit performed (yyyy-mm-dd): ☐☐☐☐-☐☐-☐☐

Auditor name: _____

Audit company: _____

GENERAL BUILDING INFORMATION

A1 Number of buildings on the property that are to be audited?   ☐☐☐

A2 Building year?   ☐☐☐☐

A3 Has there been any major retrofit/conversion/extension?
    1. ☐ Yes   2. ☐ No
       Year of major retrofit/conversion/extension?   ☐☐☐☐   ☐ Has not been performed

A4 Building location:
    1. ☐ Urban area, city
    2. ☐ Suburb, smaller urban areas with mainly multi-family buildings
    3. ☐ Detached building area, terrace house area
    4. ☐ Rural area

A5 What type of building is being audited?
    Single-family units
    1. ☐ Detached house
    2. ☐ Terrace house
    3. ☐ Semi-detached house
    4. ☐ Semi-detached two-family house
    5. ☐ Detached two-family house
    6. ☐ Other: _____

Multi-family units
7. ☐ Slab block
8. ☐ Tower block
9. ☐ Balcony access block
10. ☐ Other: _____

A6 How many storeys are there in the building?　　　　　Number of storeys ☐☐

How many storeys in the house are heated to at least 16°C during　　Number of storeys ☐☐
the winter season?

How many storeys in the house are cooled to at least 21°C during　　Number of storeys ☐☐
the summer season?

A7 How is the building utilized?
1. ☐ Residence only
2. ☐ Residence and for commercial/industrial activities

A8 Give for the building (*plan drawings or ocular survey*)
a) Circumference (external dimension)　　☐☐☐☐ m
b) Longest length (external dimension)　　☐☐☐☐ m
c) Envelope area (external dimension)　　☐☐☐☐ m²
d) Total floor area (internal dimension)　　☐☐☐☐ m²
   whereof
e) Heated floor area (internal dimension)　　☐☐☐☐ m²
   whereof
f) Residential heated floor area (internal dimension)　　☐☐☐☐ m²
g) Total volume (external dimensions)　　☐☐☐☐ m³

A9 How many apartments are there in the building?　　☐☐☐☐ apartments

What is the distribution of apartments?
☐☐☐☐ apartments　　　　≤1 room and kitchen (r a k)
☐☐☐☐ apartments　　　　2 r a k
☐☐☐☐ apartments　　　　3 r a k
☐☐☐☐ apartments　　　　4 r a k
☐☐☐☐ apartments　　　　≥5 r a k

A10 Which is the main ventilation system type? (*informative*)
1. ☐ Natural ventilation system
2. ☐ Controlled natural ventilation system
3. ☐ Exhaust ventilation system
4. ☐ Balanced ventilation system
5. ☐ Balanced ventilation system with heat recovery

A11 Are there any recordings from the most recent tuning of the main ventilation system?
1. ☐ Yes　　　2. ☐ No

If single-family house, go to A14

A12 Which storey is the apartment in?
1. ☐ Bottom
2. ☐ Intermediate
3. ☐ Top

A13 Is it a gable apartment?
1. ☐ Yes　　　2. ☐ No

A14 Number of rooms excluding the kitchen?　　☐☐
whereof bathrooms　　☐

Windows and doors

Fill in the tables for windows and glazed parts of doors in respective sector. Sum these up and enter the total value.

A15 Sum table sector North (NW-NE)   ☐ No windows

Type/Variants	g-factor	Curtain factor	Lateral sector horizontal angle	Central sector horizontal angle	Central sector horizontal angle	Lateral sector horizontal angle	Glazed area
☐-glazed							⎿⎿⎿ m²
☐-glazed							⎿⎿⎿ m²
☐-glazed							⎿⎿⎿ m²
☐-glazed							⎿⎿⎿ m²

Sum table sector East (NE-SE)   ☐ No windows

Type/Variants	g-factor	Curtain factor	Lateral sector horizontal angle	Central sector horizontal angle	Central sector horizontal angle	Lateral sector horizontal angle	Glazed area
☐-glazed							⎿⎿⎿ m²
☐-glazed							⎿⎿⎿ m²
☐-glazed							⎿⎿⎿ m²
☐-glazed							⎿⎿⎿ m²

Sum table sector South (SW-SE)   ☐ No windows

Type/Variants	g-factor	Curtain factor	Lateral sector horizontal angle	Central sector horizontal angle	Central sector horizontal angle	Lateral sector horizontal angle	Glazed area
☐-glazed							⎿⎿⎿ m²
☐-glazed							⎿⎿⎿ m²
☐-glazed							⎿⎿⎿ m²
☐-glazed							⎿⎿⎿ m²

Sum table sector West (NW-SW)   ☐ No windows

Type/Variants	g-factor	Curtain factor	Lateral sector horizontal angle	Central sector horizontal angle	Central sector horizontal angle	Lateral sector horizontal angle	Glazed area
☐-glazed							⎿⎿⎿ m²
☐-glazed							⎿⎿⎿ m²
☐-glazed							⎿⎿⎿ m²
☐-glazed							⎿⎿⎿ m²

A16 Sum the total area of non-glazed part of external doors and balcony doors. Measure from frame-to-frame.
Total door area ⊔⊔⊔ m²

RESIDENTS AND BEHAVIOUR (Information from the resident representative)

B1 How many person live in the apartment?
Number of adults (18 years and above) ☐☐
Number of children (0–2 years) ☐☐
Number of children (3–12 years) ☐☐
Number of teenagers (13–17 years) ☐☐

B2 How many weeks per summer season is *no one* living in the apartment? ☐☐ weeks
How many weeks per winter season is *no one* living in the apartment? ☐☐ weeks

---

B3 How many persons use the apartment more than 6 hours between hrs 08 and 18?
Persons during working days ☐☐
Persons during weekends ☐☐

B4 Is there a main heating system in the building? 1. ☐ Yes 2. ☐ No
At what temperature does the resident usually keep during the winter? (Occupant estimation)

	Temperature	Heating off
During the day when no one is at home.	☐☐°C	☐ yes
During the day when someone is at home.	☐☐°C	☐ yes
During sleeping hours.	☐☐°C	☐ yes

B5 Is there an automatic setback of the indoor temperature during the day/week at winter?
1. ☐ Yes 2. ☐ No

Set-point temperature	starting hours–ending hours
☐☐.☐	☐☐.☐☐–☐☐.☐☐
☐☐.☐	☐☐.☐☐–☐☐.☐☐
☐☐.☐	☐☐.☐☐–☐☐.☐☐

B6 Is there a cooling system in the building? 1. ☐ Yes 2. ☐ No
At what temperature does the resident usually keep during the summer? (Occupant estimation)

	Temperature	Cooling off
During the day when no one is at home.	☐☐°C	☐ yes
During the day when someone is at home.	☐☐°C	☐ yes
During sleeping hours.	☐☐°C	☐ yes

B7 Is there an automatic setback of the indoor temperature during the day/week in summer?
1. ☐ Yes 2. ☐ No

Set-point temperature	starting hours–ending hours
☐☐.☐	☐☐.☐☐–☐☐.☐☐
☐☐.☐	☐☐.☐☐–☐☐.☐☐
☐☐.☐	☐☐.☐☐–☐☐.☐☐

B8 When was the heating system turned on and off?
Turned on (mm-dd) ☐☐-☐☐　　Turned off (mm-dd) ☐☐-☐☐　　☐ Not available

When was the cooling system turned on and off?
Turned on (mm-dd) ☐☐-☐☐　　Turned off (mm-dd) ☐☐-☐☐　　☐ Not available

---

Heating

C1 How is the audited building mainly heated?
1. ☐ Individual boiler　2. ☐ District heating　3. ☐ Sub district heating/central boiler plant
4. ☐ Heat pump　5. ☐ Direct electrical heating　6. ☐ Other: _____

C2 Main heating type and rated power?

	Rated power	Age	Meter no./subscription no.
Individual boiler	☐☐☐.☐☐ kW	☐☐ years	☐☐☐☐☐☐☐☐☐☐
District heating	☐☐☐.☐☐ kW	☐☐ years	☐☐☐☐☐☐☐☐☐☐
Local district heating/grpc	☐☐☐.☐☐ kW	☐☐ years	☐☐☐☐☐☐☐☐☐☐
Heat pump	☐☐☐.☐☐ kW	☐☐ years	☐☐☐☐☐☐☐☐☐☐
Direct electrical heating	☐☐☐.☐☐ kW	☐☐ years	☐☐☐☐☐☐☐☐☐☐
Solar heating	☐☐☐.☐☐ m² collector area	☐☐ years	☐☐☐☐☐☐☐☐☐☐

C3 Are there service, tuning and operation records?   1. ☐ Yes   2. ☐ No

If yes

Individual boiler	efficiency ☐☐☐%	year ☐☐☐☐	
District heating	efficiency ☐☐☐%	year ☐☐☐☐	
Local district heating/grpc	efficiency ☐☐☐%	year ☐☐☐☐	
Heat pump	COP ☐.☐☐	year ☐☐☐☐	
Direct electrical heating	efficiency ☐☐☐%	year ☐☐☐☐	
Solar heating	efficiency ☐☐☐%	year ☐☐☐☐	

C4 *For individual boilers only.* Estimated annual energy distributions of various energy fuels (by responsible representative).

Do the boilers use more than one fuel?   1. ☐ Yes   2. ☐ No

Meter no./subscription no.

Light heating oil	☐☐☐%	☐☐☐☐☐☐☐☐☐☐
Heavy heating oil	☐☐☐%	☐☐☐☐☐☐☐☐☐☐
Natural gas	☐☐☐%	☐☐☐☐☐☐☐☐☐☐
Electricity	☐☐☐%	☐☐☐☐☐☐☐☐☐☐
Wood, sawdust, chips	☐☐☐%	☐☐☐☐☐☐☐☐☐☐
Pellets	☐☐☐%	☐☐☐☐☐☐☐☐☐☐
Coal, coke	☐☐☐%	☐☐☐☐☐☐☐☐☐☐

C5 Is there an automatic change-over between the various energy fuels?
1. ☐ Yes, temperature controls   2. ☐ Yes, timer controls   3. ☐ No

C6 Is there in the audited building other forms of supplementary heat generating sources other than heat pumps and solar collectors?

More than one alternative is possible.

Device	Frequency of use/week	Quantity/ use	Utilisation time	Quantity/ season	Included in bills
1. ☐ Fire place (wood)	_____	_____ m³	_____		☐ Yes
2. ☐ Electrical stove	_____	_____ W	_____ hrs	_____	☐ Yes
3. ☐ Wood heating stove	_____	_____ m³	_____		☐ Yes
4. ☐ Natural gas heating stove	_____	_____ l/h	_____ hrs	_____	☐ Yes
5. ☐ Bottled gas heating stove	_____	_____ l/h	_____ hrs	_____	☐ Yes
6. ☐ Kerosene heating stove	_____	_____ l/h	_____ hrs	_____	☐ Yes
7. ☐ The cooker stove	_____	_____ l/h,W	_____ hrs	_____	☐ Yes
8. ☐ Portable heaters	_____	_____ W	_____ hrs	_____	☐ Yes
9. ☐ Other: _____	_____	_____	_____ hrs	_____	☐ Yes
10. ☐ None					

Heat pumps

C7 Does one or more heat pump exist?   1. ☐ Yes   2. ☐ No

C8 Are these used for

Space heating	1. ☐
Hot-water heating	2. ☐
Space and hot-water heating	3. ☐
Space cooling	4. ☐
Space cooling and hot-water heating	5. ☐

C9 Are values on supplied electricity (or natural gas) to the heat pump available?
   1. ☐ Yes   2. ☐ No   Meter number ☐☐☐☐☐☐☐☐☐☐☐☐

Distribution of heat

C10 Which is the type of main heat distribution in the building?
   1. ☐ Water borne   2. ☐ Air borne   3. ☐ Electrical   4. ☐ None

---

C11 Which year was the last occasion that the heating system was tuned?
   Year: ☐☐☐☐   ☐ Unknown

C12 If available from recent records, or possibility of present readings

	Shunt group 1	Shunt group 2	
Temperature of supply heat carrier?	☐☐☐ °C	☐☐☐ °C	
Temperature of return heat carrier?	☐☐☐ °C	☐☐☐ °C	
Temperature of supply heat carrier at an outdoor temperature of 0°C?	☐☐☐ °C	☐☐☐ °C	☐ Unknown
Pressure drop in circuit?	☐☐☐ kPa	☐☐☐ kPa	☐ Unknown
Flow in circuit?	☐☐☐ l/min	☐☐☐ l/min	☐ Unknown

C13 What is the main type of heaters in the building?
   1. ☐ Radiators   2. ☐ Convectors   3. ☐ Floor heating   4. ☐ Ceiling heating
   5. ☐ Other: _____

DISTRICT HEATING

C14 Does the building have its own meter for district heating, with no other buildings included?
   1. ☐ Yes   2. ☐ No

C15 Are there service, tuning and operation records?   1. ☐ Yes   2. ☐ No
   If yes
   District heating   efficiency ☐☐☐%   year ☐☐☐☐

COOLING

D1 How is the audited building mainly cooled? (more than one alternative possible)
   1. ☐ Local units   2. ☐ Central unit   3. ☐ District
   4. ☐ Fans (without cooling)   5. ☐ Other: _____   6. ☐ No cooling

---

D2 What is the main type of cooling equipment? (more than one alternative possible)

Type	Rated power	Age	"Fuel"	
Local units	☐☐☐.☐☐ kW	☐☐ years	☐ Electricity	☐ Natural gas
Central cooling	☐☐☐.☐☐ kW	☐☐ years	☐ Electricity	☐ Natural gas
District cooling	☐☐☐.☐☐ kW	☐☐ years		
Fans	☐☐☐.☐☐ kW	☐☐ years		

D3 Are there service, tuning, operation records or rated performance?   1. ☐ Yes   2. ☐ No
   If yes

Local units	mean efficiency ☐.☐	year ☐☐☐☐
Central cooling	efficiency ☐.☐	year ☐☐☐☐
District cooling	efficiency ☐☐☐%	year ☐☐☐☐

Domestic hot water

E1 Is there a central domestic hot-water boiler?
   1. ☐ Yes   2. ☐ No, separate boilers in every apartment/compartment
   3. ☐ Hot water not available

E2 Are meters for the hot-water consumption for the building available?   1. ☐ Yes   2. ☐ No

E3 How is the heating of domestic water mainly done during the summer and the winter season?

	Summer	Winter
Water heating with conventional boiler/heat exchanger/ heat pump that also provides space heating	1. ☐	1. ☐
Heat pump only for water heating purpose	2. ☐	2. ☐
Separate electrically heated water boiler	3. ☐	3. ☐
Solar heating	4. ☐	4. ☐
Bottled gas	5. ☐	5. ☐
Hot water shut off	6. ☐	6. ☐
Other: _____	7. ☐	7. ☐

E4 Rated heat output of separate domestic hot-water boiler?   ☐☐☐.☐☐ kW   0 ☐ Not available
    Age of separate heated water boiler?   ☐☐ years   0 ☐ Not available

E5 Rated heat output of central domestic hot-water boiler?   ☐☐☐.☐☐ kW   0 ☐ Not available

E6 What is the set-point temperature for the hot water?   ☐☐☐ °C   0 ☐ Not available

E7 Size of the hot-water tank/accumulator tank?   ☐☐☐☐ m³   0 ☐ Not available

Appliances

F1 What type of fuel does the stove use?
    1. ☐ Electricity   2. ☐ Natural gas   3. ☐ Bottled gas   4. ☐ Other

    What type of fuel does the oven use?
    1. ☐ Electricity   2. ☐ Natural gas   3. ☐ Bottled gas   4. ☐ Other

F2 How many hot meals are on an average cooked each day at home?
    1. ☐ Three or more   2. ☐ Two   3. ☐ One   4. ☐ A few times per week

Electricity meter information

G1 Does the building have its own electricity meter, with no other buildings included?
    1. ☐ Yes   2. ☐ No

G2 Which electrical meters exist for the building?
    Meter 1   Meter 2   Meter 3
    ☐☐☐☐☐☐☐☐☐   ☐☐☐☐☐☐☐☐☐   ☐☐☐☐☐☐☐☐☐

G3 The meter reads the following (more than one alternative possible)

	Meter 1	Meter 2	Meter 3
Household electricity	1. ☐	1. ☐	1. ☐
Electrical space heating	2. ☐	2. ☐	2. ☐
Electrical space cooling	3. ☐	3. ☐	3. ☐
Hot water boiler	4. ☐	4. ☐	4. ☐
External lighting	5. ☐	5. ☐	5. ☐
Other electricity use on the property	6. ☐	6. ☐	6. ☐

Other electricity use on the property

G4 Are there other electrical devices or buildings on the property which use the same electricity meter?
    1. ☐ Yes   2. ☐ No?

G5 Describe the object here and write, if possible, rated power and run time as a percentage for the seasons.

	Rated power	Run time winter	Run time summer	Meter/subscription no.
Exterior lighting	☐☐☐ kW	☐☐☐ %	☐☐☐ %	☐☐☐☐☐☐☐☐☐☐
Gutter/downpipes/water electrical cables	☐☐☐ kW	☐☐☐ %	☐☐☐ %	☐☐☐☐☐☐☐☐☐☐
Car heater	☐☐☐ kW	☐☐☐ %	☐☐☐ %	☐☐☐☐☐☐☐☐☐☐
Pool heaters and pumps	☐☐☐ kW	☐☐☐ %	☐☐☐ %	☐☐☐☐☐☐☐☐☐☐
Exterior sauna	☐☐☐ kW	☐☐☐ %	☐☐☐ %	☐☐☐☐☐☐☐☐☐☐
Infra-red heaters	☐☐☐ kW	☐☐☐ %	☐☐☐ %	☐☐☐☐☐☐☐☐☐☐
Outdoor grill	☐☐☐ kW	☐☐☐ %	☐☐☐ %	☐☐☐☐☐☐☐☐☐☐
_____	☐☐☐ kW	☐☐☐ %	☐☐☐ %	☐☐☐☐☐☐☐☐☐☐
_____	☐☐☐ kW	☐☐☐ %	☐☐☐ %	☐☐☐☐☐☐☐☐☐☐
_____	☐☐☐ kW	☐☐☐ %	☐☐☐ %	☐☐☐☐☐☐☐☐☐☐

SUPPLIERS

Energy supplier name, Address	Subscript/account	Meter number	Fuel	Unit	Estimate Exact	Conv. factor
1						
2						
3						
4						

Water supplier name, Address	Subscript/account	Meter number	Fuel	Unit	Estimate Exact	
			water			

No.	Date	Quantity	Date	Quantity	Date	Quantity	Date	Quantity	Date	Quantity

APPENDIX 2

Energy transmittance by glazing and shading factors

The information in Appendix 2 is to a large extent derived from EN 832. The quantity of data has been reduced for practical purposes: the inaccuracy in the results of the fieldwork carried out by auditors must be minimized.

TOTAL SOLAR ENERGY TRANSMITTANCE FOR GLAZING

Energy transmission through transparent surfaces depends on the type of glass and the coatings. The total solar energy transmittance defined in EN 410 is calculated for solar radiation perpendicular to the glazing, $g_\perp$. Some indicative values are provided in Table A2.1.

For monthly calculations, a value averaged over all angles of incidence is required. The factor F_w, depending on type of glass, latitude, climate and orientation, is given approximately by

$$F_w \approx 0.7 - 0.9 \tag{A2.1}$$

SHADING FACTORS
Shading from an external horizon

The total viewing angle from a glazed surface is 180°; projected onto a horizontal surface, see the left-hand side of Figure A2.1. This angle is divided into four equal parts: two central sections and two lateral sections:

1. For each section, the average angle α subtended by the obstacles from the horizontal plane is determined, as shown on the right-hand side of Figure A2.1.

Table A2.1 Typical total solar energy transmittances for the two most common types of glazing. These are values for normal incidence, assuming a clean surface. For other types of glazing, use certified values or national default values

Glazing type	$g_\perp$
Single glazing	0.85
Double glazing, clear glass	0.75

2. The corresponding partial shading factor s_i is obtained from Table A2.2.
3. The shading factor s of the element under consideration is given by:

$$s = \min \left[1, \sum_{i=1}^{4} s_i \right] \tag{A2.2}$$

The sum is over the four sections.

Shading from overhangs

Please refer to section G 2.2 of EN832.

Curtain factors

The curtain factor is the ratio of the average solar energy entering the building with curtains to the energy that could enter the building without the curtains. Some values are given in Table A2.3 for curtains placed inside and outside the window.

Table A2.2 Partial shading factors s_i for external obstacles

Orientation of the façade	South (SE to SW) or North (NE to NW) azimuth		East (SE to NE) or West (SW to NW) azimuth		
Average angle α for each section in degrees	For both lateral sections	For both central sections	For the lateral section towards south	For the central section towards south	For the two other sections
0–9	0.00	0.05	0.05	0.10	0.05
10–14	0.05	0.10	0.10	0.15	0.05
15–19	0.05	0.15	0.15	0.20	0.10
20–24*	0.10	0.20	0.20	0.25	0.10
25–29	0.1	0.25	0.25	0.30	0.10
30–34	0.10	0.30	0.25	0.35	0.15
35–44	0.15	0.35	0.30	0.40	0.20
45–90	0.15	0.40	0.30	0.45	0.25

*These have been slightly modified from the original values of EN832. Shading factors are valid only between latitudes of 40° and 50°.

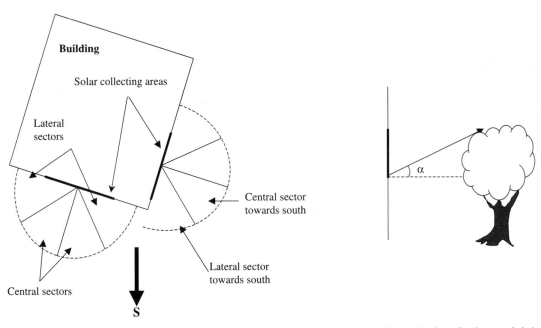

Figure A2.1 Left-hand side: a building as seen from above, where the view, as seen from behind each solar collecting area (window), is projected onto the horizontal plane. Each viewing plane is divided into four sectors. Right-hand side: illustration of the angle α

Table A2.3 Curtain factors for some types of curtains, installed inside or outside the window

Curtain type	Optical properties		Curtain factor with	
	Absorption	Transmission	Curtain inside	Curtain outside
White venetian	0.1	0.05	0.25	0.10
blinds		0.1	0.30	0.15
		0.3	0.45	0.35
White curtains	0.1	0.5	0.65	0.55
		0.7	0.80	0.75
		0.9	0.95	0.95
Coloured	0.3	0.1	0.42	0.17
textiles		0.3	0.57	0.37
		0.5	0.77	0.57
Aluminium-	0.2	0.05	0.20	0.08
coated textiles				

Estimated average fuel combustion efficiency of common heating appliances

Values of heating equipment efficiency given in Table A3.1.

Table A3.1 Values of heating equipment efficiency

Type	Heating equipment efficiency (%)
Coal (bituminous) central heating, hand-fired	45.0
Central heating, stoker-fired	60.0
Water heating, pot stove (50 gal./227.3 litre)	14.5
Oil high-efficiency central heating	89.0
Typical central heating	78.0
Water heater (50 gal./227.3 litre)	59.5
Gas high-efficiency central heating	92.0
Typical central heating	82.0
Room heater, unvented	91.0
Room heater, vented	78.0
Water heater (50 gal./227.3 litre)	62.0
Electricity central heating, resistance	97.0
Central heating, heat pump	200+
Ground-source heat pump	300+
Water heaters (50 gal./227.3 litre)	97.0
Wood and pellet Franklin stoves	30.0–40.0
Stoves with circulating fans	40.0–70.0
Catalytic stoves	65.0–75.0
Pellet stoves	85.0–95.0

Index

total energy use, annual 51, 52
total supplied energy 71
tracer gas 8, 10, *10*, 15, 16, 22
transmission factor 51, 132
transmission heat loss coefficient *see* UA values
transmission losses 5, 16–18, *17*, 19, *19*, 41, 61, 114
   North Europe 21
TRY (test reference year) 65
typical meteorological year (TMY) 65

U values 2, 15–16, 23
UA and gA method 42
UA values 1, 9, 19–21, *20*, 23
UK (United Kingdom), energy rating ii
uncertainties 10, 14, 23, 34–5, 39, 62, 116, 118
university projects, Sweden 1–6, *2*, *3*, *4*, *5*
unoccupied buildings, monitoring 16–19, *16*, *17*, *18*, 20–1, *21*
UP1 university project 2–3, *2*, *3*, 4
UP2 university project 3–6, *4*, *5*

USA (United States of America) ii, 53
   see also PSTAR; STEM
utilization factors iii, 60, 62, 68
   see also MUF method

variable of performance 71
ventilation i, 3, 54, 110
   by opening windows 99, 118, 119
   losses by 5, 9, 16, 17–18, *17*, 19, *19*, 41, 61
   rates 8, 9, 10, *10*, 16, 22, 51
VHL (Virtual Housing Laboratory) 7

water consumption 50, 55, 87, 91
water heating *see* hot water
water vapour balance 16, 17
windows i, 51, 54
   see also glazing; solar apertures; U values
wood 88
   consumption 86, 87, *89*, 109, 110–11, *110*, 113